The Social Behavior of Older Animals

The SOCIAL BEHAVIOR *of* OLDER ANIMALS

Anne Innis Dagg

The Johns Hopkins University Press

Baltimore

© 2009 The Johns Hopkins University Press
All rights reserved. Published 2009
Printed in the United States of America on acid-free paper
9 8 7 6 5 4 3 2 1

The Johns Hopkins University Press
2715 North Charles Street
Baltimore, Maryland 21218-4363
www.press.jhu.edu

Library of Congress Cataloging-in-Publication Data
Dagg, Anne Innis.
The social behavior of older animals / Anne Innis Dagg.
        p.   cm.
Includes bibliographical references and index.
ISBN-13: 978-0-8018-9050-5 (hardcover : alk. paper)
ISBN-10: 0-8018-9050-0 (hardcover : alk. paper)
1. Social behavior in animals.   I. Title.
QL775.D34 2008
591.56—dc22   2008010643

A catalog record for this book is available from the British Library.

*Special discounts are available for bulk purchases of this book. For more information,
please contact Special Sales at 410-516-6936 or specialsales@press.jhu.edu.*

The Johns Hopkins University Press uses environmentally friendly book materials,
including recycled text paper that is composed of at least 30 percent post-consumer
waste, whenever possible. All of our book papers are acid-free, and our jackets and
covers are printed on paper with recycled content.

To my two oldest friends

Mary F. Williamson

Rosemary A. Rowe

CONTENTS

ACKNOWLEDGMENTS

I WOULD LIKE TO THANK all those who helped me in the creation of this book by reading short sections of the manuscript or by sending me information about published research or particular old animals. They include Alan Cairns, Wendy Campbell, Bernice Grant, Heidi Libesman, Bev Sawyer, Elaine Sim, and Anne Zeller. I am grateful also for the fine local libraries and librarians—the Trellis tri-university system of the University of Waterloo, Kitchener Public Library, and Waterloo Public Library. It has been difficult to collect information for this book, so my especial gratitude goes to the researchers and authors who included an index in their books with entries for "old age." You are few and far between. Thank you.

For those at the Johns Hopkins University Press, my heartfelt thanks to Dr. Vincent Burke, senior editor, for his deeply appreciated encouragement and wise suggestions about organizing the book's material; Kathleen Capels for her meticulous copyediting of each chapter and a helpful insight into the orca Eve's behavior; Deborah Bors, for her useful advice and the oversight of the final preparation of my book; and Robin Rennison and Brendan Coyne for their efficient and pleasant communications with me.

The Social Behavior of Older Animals

Introduction

THIS BOOK IS ABOUT the social behavior of mammals and birds well past their prime who live either in the wild or in captivity where they have large areas in which to move and interact with others. It does not include data from animals kept in small cages and subject to experimentation in research facilities.

The question of who is old is fairly easily answered for people. Oldsters who take advantage of seniors' discounts are rarely asked to produce identification to prove their age. They often have gray or white hair, and their bodies provide other physical clues.

What about other animals, where signs of aging such as wrinkles and spots are covered with fur or feathers? By definition, older animals are nearing the end of their lives, but that does not mean we can distinguish them from their companions. Indeed, most behavioral research on wild animals does not mention older individuals at all. When colleagues and I published scientific books on giraffe (Dagg and Foster 1976) and on camels (Gauthier-Pilters and Dagg 1981), using data we had collected over years in the wild and everything about these species published in the literature, we had almost nothing to report about older animals—and at the time, we never noticed this deficit.

Struhsaker (1975), in his extensive research on red colobus monkey behavior, differentiated individuals at times into young infant, infant, old infant, young juvenile, juvenile, old juvenile, subadult, approximate adult, and adult. But he had no category for "old adult."

Nor have non-scientists been interested in the topic. In 2005, in a small newsletter read by 400 seniors, I asked if anyone could give me information about the behavior of older animals. Many of them had companion pets who were as gray as they were, but no one offered any anecdotes.

One problem was that, until recently, people believed that wild animals did not live to be old, dying instead from accidents or disease, or being killed and eaten by predators (Hrdy 1981). Therefore, there is little information available about aged individuals in older books and articles. The well-known naturalist Ernest Seton Thompson (who later called himself Thompson Seton), in his famous book *Wild Animals I Have Known* (1942), wrote that "the life of a wild animal *always has a tragic end*," and that "no wild animal dies of old age."

We now have the amazing power to collect information using Internet search engines such as Google, but searching there with the words "old animals" or "aged animals" brings up millions of items, including information on the Old World, Old English, kittens aged 3 months, and aged cheese.

Information used in this study came from a large number of books and articles on animals, the most profitable being those written by zookeepers, wildlife zoologists, and individuals or groups of people who study or simply love animals. It was not possible to merely skim the books, because words such as "old age" or "aged animals" are seldom included in the index. To ensure that I had up-to-date information, I searched article titles for pertinent data published between 2000 and mid-2007 in issues of the following academic journals: *American Journal of Animal Ecology, American Journal of Human Biology, American Journal of Physical Anthropology, American Journal of Primatology, American Naturalist, Animal Behaviour, Behavioral Ecology and Sociobiology, Behaviour, Canadian Journal of Zoology, Current Biology, Ecology, Ecological Monographs, International Journal of Primatology, Journal of Mammalogy, Nature, Oecologia, Primates, Quarterly Review of Biology,* and *Science.*

THIS INTRODUCTION FIRST CONSIDERS general knowledge about elderly animals, and then the bias inherent in collecting information on their social behavior. Throughout, I often refer to wild animals as "older" rather than "old," to reflect the difficulty of knowing exactly how old an older animal is.

Ten Basic Facts about Aging in Animals

1. Animals in zoos tend to live longer than wild ones, because they do not have to worry about predation or lack of food or water. However, this is not true for large mammals such as orcas (killer whales) or elephants—

the latter can live for 60 or 70 years in the wild, but they usually die by their forties in captivity (Gorman 2006).

2. Older animals are slower and less agile than their junior selves, and may suffer from arthritis, diabetes, cancer, heart disease, and other health problems, including mental confusion. As John Grogan (2005) stated in describing his ancient dog Marley, he and many other old animals are not only deaf and have poor vision, but their fur or hair is falling out, they are incontinent, they have arthritic hips that make it hard to get up or lie down, their teeth are broken or rotted, their confidence is eroded, and they snooze a lot. Old nonhuman animals may suffer from osteoporosis. One elderly female chimpanzee had a bone mineral density below what would indicate osteoporosis in humans; it was surmised that only her posture, locomotion, and trunk-sacral anatomy prevented her bones from fracturing (Gunji et al. 2003). Older people may suffer from these same problems, but not all of them will; women athletes in their fifties are now beating those in their twenties in long-distance running. The younger women are more physically fit, but the older ones have a mindset that involves not only physical but also emotional and spiritual energy (Kolata 2007).

3. Older mammals are often gaunt, with fur or hair that has turned gray or white. Aging leopards tend to have faded color spots (Bailey 1993), but older giraffe may have spots that are darker than those of animals in their prime (Dagg 1983). Elderly birds usually have plumage similar to that of young adults; for example, an old (at age 11) cliff swallow looked just like the other adult swallows (Brown 1998).

4. Males and females of a species may have very different average life-spans. Orca males, for example, have a longevity of about 40 years, while females often live into their fifties (Knudtson 1996).

5. Healthy female mammals are getting old if they produce fewer young than formerly (reproductive senescence), and are definitely in that category when they stop reproducing entirely. However, exceptions are possible. One Asian elephant, Tara, became pregnant in her mid-sixties, a decade after most elephants stop reproducing (Chadwick 1992). One aging lion in Nairobi National Park was assumed to be sterile because she had had no cubs for two years, but then she bore a litter (Schaller 1972a). Laboratory data on hundreds of older rhesus monkey females show that all experienced a decrease in fertility with age, but few reached a true

menopausal state (meaning a permanent cessation of ovulation) comparable to that in women (Small 1984). In contrast to mammals and birds, many female fish, amphibia, and reptiles continue to grow during their lifetimes, with their increased body size allowing them to produce more young in each successive year (Clutton-Brock 1984). Females of most species reproduce until they die.

6. Healthy males may stop reproducing when they grow old. One ancient wolf in the wild had small testes that were no longer producing sperm (Mech 1996). Jan Smuts, a long-lived male giraffe at the Taronga Zoo in Sydney, Australia—who looked ancient with a gaunt neck and face and almost black spots—no longer mated with the females around him, so a younger male, Oygle, had taken over this function (Dagg 1970).

7. Small animals in general have a far shorter lifespan than large animals, and longevity is correlated to some extent with metabolism. An elephant who lives for perhaps 65 years has a normal heart rate of less than 25 beats per minute, while a masked shrew who lives only a year or two has a rate of about 1300 beats per minute (Gunderson 1976). Insectivorous bats also have a fast heart rate while active, but their lifespan is about 10 times as long as that of shrews, because they spend much of their existence in hibernation (Barclay and Harder 2003).

8. Longevity is an adaptive trait, positively correlated with unique attributes such as flying ability (sometimes), possession of armor (in turtles and armadillos), and life underground (for moles and mole rats) (Carey 2003). Parrots may live for 50 years, while tortoises can survive to about age 175—the tortoise Harriet from the Galápagos Islands, who died recently, was apparently taken to England by Charles Darwin and later sent to Australia, where the climate was more suitable (Rook 2006; Discovery Channel 2006). The echidna, an egg-laying mammal, is primitive but also very long lived, surviving sometimes for 50 years (Crandall 1966). Human beings have an exceptionally long lifespan for their size, much longer than that of other primates.

9. In dogs, larger species such as Great Danes and collies have a short lifespan of 8 to 10 years, while small dogs live much longer on average. Among dogs weighing 9 kg (20 lb), cairn terriers have a typical lifespan of 13 to 14 years, a miniature poodle 15 to 16 years, and a Cavalier King Charles spaniel only 11 or 12 years (Coren 1994).

10. Researchers in different areas may have varying standards for old

age. In Gombe, Tanzania, Jane Goodall (1986) defined old chimpanzees as those who were at least 33, while at the nearby Mahale Mountains research area, old age was considered to be 41 or older (Huffman 1990).

For this book, I have sometimes had to use my own judgment, case by case, to decide who was old. In general, I assume animals are old if the author who writes about them says they are. (This resembles Darwin's sensible definition of a species: a species is a species if an expert on the group defines it as such. When I first heard this maxim I was horrified at the lack of rigor it seemed to imply, but now I realize that it is eminently practical. Who better to decide than a person studying the animals?)

Of course, very elderly animals near death are usually easily recognized as being old. Often their teeth are broken or gone, so they cannot eat properly; indeed, this factor often causes their death. Such animals tend to be loners. They either cannot keep up with their group as it forages (elephants), or they have been driven out of the group because of their age (hyenas). For an intra-species comparison, measurements have been taken of over 600 rhesus monkeys of all ages, showing age-related changes in their vision, muscles, and bones (DeRousseau et al. 1986).

Biased Information about Social Behavior

HUMANS KILLING OLDER ANIMALS

Joyce Poole has written a fascinating book, *Coming of Age with Elephants* (1996), about growing up in Africa and beginning research on elephants in Kenya when she was only 19 years old. She spent thousands of days driving around the Amboseli National Park, observing wild elephants, especially males, and writing down what they were doing. Within a few years, by using individual photographs and computer cards, she and Cynthia Moss, her mentor, could individually identify all the elephants in the park—164 adult males and 451 adult females and calves—by such characteristics as size, tusk shape, and ear descriptions (involving their size, notches, tears, and holes). Altogether, Poole spent 14 years in a tented camp in the park, studying male elephants both for her graduate work and as part of the Amboseli Elephant Project. By 1995, she had increased her male count to 450 animals.

Wow! I thought. This will be a gold mine of information on the behavior of old elephants. Poole and Moss must have observed scores of them over the years. I turned to my copy of Poole's book with enthusiasm. Alas, my excitement was misplaced. When Poole began her research in 1975, poaching in Amboseli Park was extensive, with the largest and oldest animals being slaughtered for the ivory of their huge tusks. As she checked through the photographs of 68 known males, she realized that such behemoths were no longer around. Although the lifespan of adult elephants in the wild is about 65 years, no elephant in the park even began to approach this age. Of the 68 males, only eight remained who were over 30 years old, and only one, Iain (M13), was over 40. By 1995, only six of this original group were still alive.

What about the behavior of older males? Are they patriarchs? Are they important in elephant society? Unfortunately, we do not have enough data to answer such questions; largely because of culling and poaching for ivory, many African countries no longer have *any* older males and, therefore, no natural populations of elephants. Even if poaching and culling are curtailed, wherever elephant herds roam near agricultural lands there are continuing problems (and killings), largely attributable to older elephants, who are the most experienced and boldest in foraging and trampling agricultural crops.

In Zaire, killing elephants for their ivory was banned in 1977, but it has been replaced by killing elephants because they are raiding crops, even in areas where there are no crops. A recent inventory found ivory from 6500 animals, with older elephants as the preferred target, because of their large tusks (Chadwick 1992).

In the Central African Republic, elephants have been killed for their ivory for years, the oldest—and therefore largest individuals, of course—first. A 1986 inventory of thousands of tusks, obtained from elephants who had been shot for allegedly raiding crops, turned up none from animals older than 35. Since older, experienced elephants are the most notorious crop raiders, it seems as though the entire country has few or no elephants beyond middle age.

The same problem of excessive harvesting exists for whales. Before commercial whaling began, large whales lived into their sixties and seventies, while smaller bowhead and sperm whales may have survived much longer than this (Whitehead 2003). Since then, as killing technologies have improved, whalers have been slaughtering their prey more and more effectively, by the thousands and even millions. They only largely ceased doing so in 1986, with the worldwide moratorium on commercial whaling. Wherever possible, the targeted whales were the largest, most lucrative animals, so that even today, whale populations remain skewed toward younger individuals. (Some nations, especially Japan and Norway, continue to slaughter whales despite international agreements not to do so, claiming that this is necessary for "scientific research" [Glavin 2006].) The behavior of current whale groups almost certainly has been affected because of this harvesting, but we do not know how. The natural behavior of many of these large mammals will probably never be fully known, given our human insistence in meddling with their populations.

TEN PROBLEMS WITH RESOURCES

There are many other reasons why information about the social behavior of older animals has been biased, as well as little observed, reported, or understood.

1. Few methods have been developed to determine the age of animals that have no economic, scientific, or recreational interest for people. Veterinary books dealing with livestock do not include information pertinent only to older animals, because virtually no stock individual lives out its full lifespan. Animals raised for food are killed in early adulthood, and racehorses are not allowed to race in North America beyond their fifteenth birthday (Delean 2007).

2. Most animal species have not been studied by field ethologists, so we have little or no idea about their behavior in general, let alone that of their oldest members.

3. Any animal population tends to have only a few old individuals. For example, among the chimpanzees studied by Jane Goodall, roughly 12 percent were considered old at any one time, and during 30 years of research, only three females and six males appeared "ancient looking" (Engel 2002).

4. Virtually nothing is known about the social behavior of members of non-social species, even those that have been studied, because many individuals within these species seldom interact.

5. For some species, old age correlates in part with an individual's usually unknown reproductive history. For the willow tit, females who breed every year grow old and die sooner than those who skip breeding early in

their lifetimes (Orell and Belda 2002). Similarly, female red-billed choughs who lay small clutches and fledge few offspring early on have longer lives than females who are more productive when they first mature (Reid et al. 2003).

6. Few small mammals in the wild live to an old age, and their behavior is almost impossible to observe (for example, for white-footed mice, see Havelka and Millar 2004). Yet laboratory experiments indicate that there are behavioral changes with age. For example, old rats in captivity have decreased motor activity compared to young rats, and also differ from them in their emotional and social behavior (Boguszewski and Zagrodzka 2002). Unlike mammals, birds can be aged when they are banded or ringed as nestlings, and later captured in a mist net (see chapter 2). About 200 million birds have been banded worldwide, although few such marked individuals are subsequently recaptured (Berthold 1996).

7. From necessity, much of the information in this book is anecdotal. Yet observations about one old individual do not necessarily apply to other individuals of the same species. Each group has a history, as well. One can watch what individual baboons do, and assume that this represents baboon social behavior, but one must also know what has gone on in the troop's past, even the remote past, which is still reflected in the actions of present-day animals (Sapolsky 2001).

8. Numerous toxic chemicals from human pollution have accumulated in the bodies of predators, especially those who are older. Three orcas dying in the Puget Sound area off the west coast of North America had PCB levels of 250, 370, and 661 parts per million; the U.S. FDA standard for PCBs in fish for human consumption is 2 parts per million (Lord 2004). Many endocrine-disrupting chemicals now present in the environment are known to cause bizarre behaviors in wild animals (Clotfelter et al. 2004).

9. Depending on the species, aging methods are not always accurate, and may reflect such things as diet (which affects wear on teeth) or extreme weather conditions. Female waterbuck in Uganda seldom live more than 12 years, but one female with exceptionally strong teeth survived until age 18 (Spinage 1982). In contrast, the elderly chimpanzee Flo's teeth were worn down to her gums eight years before her death (Goodall 1986).

10. It is usually impossible to tell how old adult animals are in the wild, unless some method to age them, such as tagging youngsters or counting

the annual rings on teeth or horns, has been devised by wildlife experts. In the baboon troop observed by Barbara Smuts (1985), it was impossible to tell an animal's age from his or her physical appearance, especially for males. In 1977, Boz had been an adult male for one year, while Alexander and Sherlock were both subadults (about 8 years old), nearing full size. Six years later, Boz looked much the same as he had earlier, still about 24 kg (53 lb) in weight, with a rich mantle of hair over his neck and shoulders; by contrast, the other two seemed old, with slow and labored movements, drooping lower jaws, and scarred faces. Both had badly worn or broken canine teeth. (Boz and Alexander would become good buddies, as described in chapter 10.)

Most important of all, we must remember that zoologists studying animal behavior are usually trying to correlate the production of young with evolutionary concerns. Who has the most healthy young? Great. Are more young produced by some individuals with one type of characteristic rather than another? Wonderful. What about animals who no longer reproduce? Forget them.

But let us not forget them.

Scientists studying the behavior of animals now largely agree that they have feelings and emotions, just as people do. As Marc Bekoff (2006) stated: "There are pleasure-seeking iguanas, amorous whales, elephants who suffer from post-traumatic stress disorder, pissed off baboons who beat the stuffing out of others, sentient fish and a sighted dog who serves as a Seeing Eye dog for his canine buddy." This is why, of course, we cannot assume that the behavior of one animal is representative of its entire species. Ethically, it also means we should think of animals as the sentient beings that they are, and it helps if we use the names of individual animals when these are known. If possible, in the text I refer to individuals as "he" or "she" rather than "it," and use "who" or "whom" to denote them.

CHAPTER I

Evolutionary Matters

BEHAVIORAL ZOOLOGISTS ANALYZE the behavior and adaptations of animals to determine why and how these characteristics developed over time. The reason they evolved was to improve the species' potential to reproduce. The more offspring individuals produce compared to their peers, the more likely they are to have their genes survive. Yet for most mammals and birds, older individuals are either no longer reproducing, or reproducing at a slower rate than younger adults. Given that reproduction is the bedrock of evolutionary theory, can the lives of social animals who have reached an advanced age be important in evolution? Surely, from an evolutionary perspective, the wiser course would be death rather than an old age where consumption seems to exceed contribution. But since animals *do* live to be old, there must be an evolutionary reason for this.

Older social animals are vital to their group for two main reasons: (1) they have a good genetic inheritance that they pass on to their descendants, and (2) they have extensive experience with their environment and their species' culture, both of which they share with the younger members of their group. These aspects of their existence, inextricably intertwined, are considered under four headings in this chapter. First, to have survived to old age, senior animals must have good genes in general, including genes involved with a successful personality; as it is in human beings, personality in wild animals is likely to be most fully developed in adults and elders. Older animals also have a vast store of knowledge to impart to younger group members, which is vital in helping their whole group survive and thrive during hard times. Second, some species have females who live fairly long post-reproductive lives, indicating that these elders must be important for some reason. Why else would they be using up resources that could benefit younger breeding animals? Third, some genes are tailored to affect the behavior of aging animals, with positive results for the group as a whole. Fourth, many older animals display nepotism or altruism that can improve the lives of their younger relatives, thereby also enhancing their own genetic heritage.

Older Animals as Success Stories

Older animals are, by definition, winners, because most of their peers have long since died. They have good genes.[1] Over the years, they have survived all the perils that afflict their species—accidents, famine, drought, aggression from within their group, and predation. They may no longer be interested in reproducing, but they usually have a large number of progeny to their credit. They are role models and perhaps mentors to both their own young and those of their peers.

Some older individuals within social species are especially important evolutionarily (personality is genetic in large part), because they have been far more successful than other elders in passing along their genes to succeeding generations. Species evolve because, over time, the DNA of certain individuals alters in a way that makes them more adaptable to their changing environment. They produce more successful progeny than others, so that their DNA becomes predominant—they are exemplars of the future for their species. For example, when Beethoven, the beloved male gorilla studied by Dian Fossey (1983), died of old age, he had sired at least 19 offspring; later his sons Icarus and then Ziz and Pablo took over this function. Many of Beethoven's progeny left the troop, thus widely spreading his genetic inheritance. Like Beethoven, Effie, the top female in his troop, also had a large genetic influence. When she died of natural causes, she left behind at least seven living offspring. These, in turn, produced young of their own, who continued to live and reproduce.

Another amazingly successful female was Number 9, a Canadian wolf who, along with 30 others, was introduced in 1995 and 1996 into Yellowstone National Park, where wolves (*Canis lupus*) had earlier been exterminated (Smith and Ferguson 2005). In 1995, while in an acclimation pen waiting to be released into the park, she mated with a stranger, Number 10, much to the joy of the zoologists, who had not expected such good news in the first year of the transfer. Unfortunately, this male was shot soon after his release. Number 9, alone, wandered outside the park and scratched out a depression in the soil when it came time to give birth to her eight pups. But a lone female has little chance of raising her young, since she must go off to hunt for food, leaving her family vulnerable to predators. Because of this, her human watchers swept her and the pups up by helicopter to take them back to a pen in Yellowstone. They released the wolves in the fall, when the pups were much larger. Number 9 was then fortunate to link up with a strong male, Number 8, who became an instant father to her pups. She was the alpha female in the new family pack.[2]

During the next four years, the pair successfully raised four litters; by 1999, DNA tests showed that 79 percent of all the wolves in Yellowstone were related to Number 9. In 2000, when her black fur had turned gray, she had her final litter, although no pups survived. By this time she was eight or nine years old, elderly for a wolf in the wild. Several members of the public wondered about having a statue erected to honor her remarkable contribution to the wolves of Yellowstone.

When the battery in Number 9's radio collar finally went dead after six years, when her coat was snowy white except for a touch of black on her tail, the zoologists did not recollar her. Renée Askins, in her book *Shadow Mountain* (2002), wrote that Number 9 would at last "be able to slip back into mystery, into the light and shadow from which she came, free of the meddling of our race, free from our good intentions. When the mother of the Yellowstone wolves dies, where she dies, and how, I hope we will never know; it should be that way."

Although an individual may be an exceptional breeder, this does not always mean that his or her genetic line will prosper. Jonathan Weiner, in his book *The Beak of the Finch* (1994), about Peter and Rosemary Grant's 21-year study of finches, describes Famous Bird, one of the many hundreds of banded finches known by sight. This sparrow-sized bird was 13 years old, as old as any of the thousands of finches who had lived on Daphne Major Island in the Galápagos, but Peter Grant announced, "I don't think there's a single one of his offspring flying around. Not *one* has made it to the breeding season." Famous Bird had been a father many times, but never once a grandfather. When he died at an advanced age, his genetic line died with him. Sons sing the song of their fathers, so his melody would never again be heard.

From the study of twins, the trait of personality in human beings—with their large brains—is thought to be about 20 to 50 percent genetically based (Segal 1999), with the remainder shaped by an individual's experiences and environment. In contrast, purebred dogs, bred to have specific behaviors, probably have a much greater percentage of their personality traits hard-wired. Labrador dogs are in general good-natured and friendly to children, whereas rottweilers and pit bulls are assumed to be aggressive (although individuals with good owners can also be friendly). Probably the genetic component of personality in wild animals lies between these two extremes. As Peter Steinhart (1995) wrote in his book *In the Company of Wolves,* "there are soft wolves and hard wolves, soldiers and nurses, philosophers and bullies." For old Yellowstone wolves in the same pack, Number 42 was an amiable, caring mother, while Number 40 (see chapter 11) was a vicious despot (Smith and Ferguson 2005). An animal's personality may make it either less or more successful as a parent. This means that when information about older animals is anecdotal, as it often must be here for lack of research studies, the behavior of one animal cannot be said to represent that of the species.

In solitary (but not asocial) species, older individuals may also have superior reproductive success. These include osprey (A. Poole 1989) and pumas (Logan and Sweanor 2001). Moreover, certain individuals may be exceptionally successful. A 13-year-old wild female leopard, Umfazi, who was at last becoming less efficient as a hunter in her old age, had a splendid record of reproduction, producing nine sets of cubs, of which nine individuals survived to become independent (Hinde 1992).

Importance of Post-Reproductive Females

Older animals are so important in some species (including human beings, a few other primates, and some whales) that they continue to live long after they have stopped reproducing. Because their deaths would free up resources (especially food) that could be used by their younger breeding relatives, evolutionary theory assumes that there must be a good reason for such "postponed" expirations. Presumably females who live past their own breeding periods are somehow helping their progeny to improve their lives. This "grandmother hypothesis" is considered in depth in chapter 12. Post-reproductive animals include some domestic livestock and family pets, but their lives have been too extensively manipulated by people to be considered natural (Packer et al. 1998). There is little information about post-reproductive males in the wild, since which males sired which young is not known, or if they sired any at all.

One example of the grandmother hypothesis is female pilot whales, who live for many years after they have stopped breeding. For centuries, Japanese fishermen have been driving entire schools of whales to slaughter, among them short-finned pilot whales (*Globicephala macrorhynchus*). From 1975 until 1984, 20 schools comprising a total of 717 individuals were butchered. While their bodies were being flensed, Australian Helene Marsh and Japanese scientist Toshio Kasuya teamed up to collect data on the reproductive status and age of the animals (1991). This involved studying their reproductive organs and determining their age from the growth layers of their teeth.

The data gathered for the elderly pilot whales were amazing. Of 92 females living wild in the ocean who were more than 36 years old, none was pregnant and, although females lived to be as old as 63, none over 40 were ovulating. For the entire population, about one quarter of the mature females were probably no longer breeding, an incredibly high proportion. Perhaps most of these older females were vital for remembering such things as where fish for the pod were to be found, but a significant number were decidedly important for the life and health of young whales. These elderly mothers produced milk for calves until they reached puberty, at 8 years for females and 13 for males. One female nursed young for 17 years. The number of whales who were lactating in relation to the number who were pregnant was considerably greater in females over the age of 30 than in younger females.

The researchers surmised that if a female died fairly young, several of her calves might be in jeopardy (each female only produces four or five young in her lifetime); by stopping her reproduction early, she can more likely raise these final young successfully.

In addition to this possibility, pilot whales, like sperm whales, dive deep while foraging—navy scientists have trained them to go down more than 600 m (2000 ft) on command. When mothers swim this far below the surface to obtain food, they must leave their young, who are physically unable to accompany them, at the surface (Whitehead 2003). There, the calves may suckle from post-reproductive females who are looking after them, as apparently occurs in sperm whales; these oldsters are therefore "a crucial part of pilot whale society" (Norris and Pryor 1991).

Genes Turned On in Old Age?

At present, it is impossible to know what genes may switch on and be vital in older, as opposed to younger, animals. We know, for example, that senior males from social communities, such as those of African buffalo, often end up as relatively solitary beings, even though it seems unlikely that they have been pushed out of their group or failed to keep up with its travels. Older Japanese monkeys have been commended for bringing serenity to their troop, and many elderly individuals of other species also become more relaxed and less social than when they were younger (see chapter 16). Are these changes genetic, or just weariness of spirit?

It seems to be genetic that langur females, who are not aggressive in their prime when they are raising young, turn out to be so later on when they band together to fight off danger that threatens their troop (see chapter 12). Genetics also seems to be the reason that females who are at the end of their reproductive lives invest more in their final young than they have in past offspring, as discussed in chapter 6. It makes evolutionary sense for a female to expend more resources in producing an especially heavy infant, and to spend more time nursing and tending to this youngster, if it is her last in effect, her last hurrah. What about mares over 9 years old, who are more effective in parenting their foals than are younger horses? The seniors supply similar care to their young, but they target that care to the foal's first 20 days, when its life is most at risk (Cameron et al. 2000).

Altruism, Reciprocal Altruism, and Nepotism

Altruism and nepotism remain controversial topics in biological circles. One predominant hypothesis of social evolution is kin selection, whereby an individual will tend to aid and support a relative rather than a non-relative. Nepotism occurs when an individual specifically favors his or her own close relatives. If you help someone who shares many of your genes, this may be because your aid will be advantageous in increasing the number of your shared genes within the population (Roach 2003). By contrast, an individual who offers food or support to unrelated or less-related animals seems to be aiding them at the expense of his or her own well-being, or that of his or her progeny.

An alternate hypothesis to kin selection is that helpers may increase their own future reproduction by cooperating with unrelated individuals. Do animal altruists help others because they are kin, or because they are neighbors? The second option works well for seniors who have lived long enough to have acted as altruists and reaped its possible benefits; "longevity increases the likelihood of reciprocal altruism" (Ridley et al. 2005).

Depending on the social structure of their group, altruism or nepotism may be practiced by older animals, because either can be effective in increasing the possibility that an individual's DNA is passed on to the next generation. Sarah Blaffer Hrdy (1981) asserted that the behaviors of older female langur monkeys (*Presbytis entellus*) and Japanese monkeys (*Macaca fuscata*) represented, respectively, altruistic and nepotistic behavior. As is the case with other monkey species, female langurs show their age by slowed movements, weight loss, wrinkled faces, changes in skin pigmentation, sparser fur, and worn teeth. Hrdy found that the older females, the altruists, reacted aggressively when their troop was threatened, rushing forward to defend it (see chapter 12). They sustained injuries while they were protecting younger females and infants that were not their own. The langurs usually lived in a small one-male group, so most of the young had the same father. Within this band, females dropped in rank as they grew older, while young females rose to dominance as they reached reproductive status. By acting altruistically, even though they had lost their dominant status because of their age, the older females helped all their relatives at the risk of their own health.

While Hrdy learned that older female langurs behaved with altruism, their daughters and granddaughters, the young dominant females, did not. In their own self-interest, they retired to a distance in times of danger, watching the oldsters battle on their behalf. As their mothers' peer group aged, the daughters turned nasty, sometimes driving them away, snatching food even from their mothers if they felt like it, and pushing elders out of shady resting spots so they could rest there themselves. Even so, the battle-ready older females ignored such ingratitude from the younger generation.

In contrast to the langurs' small one-male groups, Japanese monkeys live in large aggregations, where members are less likely to be related because several males will have sired the infants in the troop. Older females do not defend their troop, as elderly langur females do, but rather act nepotistically by defending only the young whom they know belong to their own direct lineage.

Results from extensive research on a colony of 11 langur females living in a facility in California (Dolhinow et al. 1979) contradict some of Hrdy's findings. This study found that old age in the females did not result in decreased dominance, and that older langur females did not have the very low reproductive value postulated by Hrdy (1977). The authors reported instead that "few female langurs survive to even close to menopausal age in the rigors of the field and that the estimates of age employed by Blaffer Hrdy are inadequate." For the Hrdy model to work, they wrote, "it is essential that the older females are really as old as 25, 26, or 27 and that they have ceased to reproduce successfully," conditions which had not yet been tested.

Pride lionesses (*Panthera leo*), who are all related, are altruistic to some extent in that they allow cubs who are not their own to suck if times are good. Lion prides permit older females—with their worn teeth, gaunt bodies, and unkempt fur—to remain a part of the group and feed at kills they have not helped bring down. In contrast, younger female lions at times can be anything but generous. One wonders if these older females have any remembrance of their own behavior when they were in their prime and refused to let their own cubs feed at kills when meat was scarce. George Schaller (1972b), who studied Serengeti lions for four years, noted that "it was depressing to see a starving cub totter to its [feeding] mother, each rib sharply outlined beneath its unkempt hide, and receive a vicious cuff instead of a bite to eat."

The altruistic behaviors described above can have evolutionary significance, but individuals, depending on their personalities and life experiences, may also behave in ways that seem to defy evolutionary theory. Among the gorillas studied by Dian Fossey (1983), Rafiki, when old, did not get along with his son Solomon (see chapter 10), in contrast to the behavior of Beethoven, who supported his offspring Icarus and Pablo to succeed him as dominant males. Rafiki harassed his son and drove him from their troop; Solomon was forced to lead a solitary life for a while before eventually joining two females to form a new troop. (Of course, such non-evolutionary behavior occurs in many human families, too).

The evolutionary concepts of altruism, reciprocal altruism, and nepotism are obviously complicated.

IN SUMMARY, THE BEHAVIOR of older social animals is related to evolution in various ways. Seniors no longer reproduce as regularly as younger females, or cease to reproduce at all, yet in many species they remain important or even vital to their groups. By collecting more information on the lives of older animals, we should begin to be able to tease out the relative impact of genetic and experiential forces on their behavior.

CHAPTER 2

Sociality, Media, and Variability

Sociality

ALL ANIMALS ARE BORN WITH INSTINCTS, which are universal in each species, that have a genetic basis. Some of these instincts are related to individuals who have reached an advanced age, as was discussed in chapter 1. Yet experience, which "relates to an animal's conscious, sensory encounters with the world" (Balcombe 2006), is of more importance to these individuals. All older animals, almost by definition, have a great deal of experience from their past that underlies their social behavior. The focus of this and subsequent chapters is on older animals whose age has affected their behavior in observed and reported ways.

First, what defines social behavior? It is predominant in social species of mammals and birds (and present in many fish and invertebrates, too, although these are not considered here). Life forms evolved billions of years ago as solitary one-celled beings who lived by genetic instinct, responding in predictable ways to internal and external stimuli. During the course of evolution, many species came to live in groups (for example, coral reefs, ants, social dinosaurs, lion prides, flocks of starlings), even though the individuals were then forced to share resources. The salient condition was that they were able to survive and reproduce more successfully in groups.

Simplistically, social living makes sense for predators if individuals can obtain such things as additional meat, or a suitable mate, more efficiently in a group than they could living alone. Whales acting as a group can corral a school of fish that will easily feed them all (see chapter 15); several lions are needed to bring down a buffalo that will provide a feast for days. For herbivores, social living is sensible because the more eyes there are, the more likely the animals are to spot predators before they can attack. Even if a predator is successful, the larger the number of prey, the less likely it is that any one individual will be killed.

These general reasons for sociality are backed up by statistical analyses. For example, sea lions breeding in Argentina have a much better chance of raising their young if they do so in groups; only 1 of 143 pups born to gregarious females died before the end of the breeding season, while 60 percent of the 57 pups born from solitary mating pairs did so (Campagna et al. 1992). Likewise, pride lions successfully raise many more young than do nomadic females (Schaller 1972a).

Largely solitary animals (who often favor well-vegetated environments where they can live mainly unobserved) also possess social behaviors—mothers live with their young until they become independent, males and females get together to reproduce, deer gather in deer yards during snowy winters. Tigers leave scent marks on bushes and claw marks on trees to communicate with other tigers (regarding their identity, sex, and reproductive condition), while male orangutans communicate their presence to other orangutans with thunderous "long calls" (Masson 2006).

Older animals are vital in long-lived social species, because they are experienced and can recall solutions to past dilemmas. This experience is imperative, as memory can mean the difference between life and death. In times of drought, an elderly elephant matriarch—one who can lead her herd to a water source she remembers from 40 years earlier, and show the young how to scoop up sand to reach moisture—can save lives that would otherwise have been lost. Similarly, experience may have taught her that people slaughter elephants, and thus influence her to keep her herd away from human activity.

Being social also means that individuals in a group have evolved to depend on their fellows. Young elephants will die if they do not have mothers and other relatives to help raise them. These elders show the watching youngsters traditional elephant ways, such as which foods the adults eat and how they interact with each other. The experience gained by older animals and passed on to younger ones is a central core of social life.

Species' Media

What zoologists have been able to find out about the behavior of a species overwhelmingly depends on what media that species inhabits.[1] Large land animals are the most fruitful to study, because they live more or less like ourselves—perhaps less, because our sense of smell is so inferior to theirs—and we can empathize with how individuals react to each other and to their environment. Behaviors based on the past experiences of much-observed and highly social elephant matriarchs are the ultimate example of the importance of older individuals (see chapter 3).

We know little about the behavior of aquatic species, young or old. We cannot see much of what these animals are up to in the water, or understand viscerally what it means to exist there—where human beings move ineptly and have to always be alert to the need to breathe, where gravity affects our body differently than in the air, and where communication sounds are carried through water. There is virtually nothing known about the behavior of older fish in the wild, so this chapter considers only our (limited) knowledge of the behavior of aged whales and dolphins. Sociality is the norm in ocean-dwelling mammals. Their environment is unforgiving—if they do not come to the surface every few minutes to breathe, every day of their long lives, they will die. Much of their behavior is thus hard-wired: the ability to swim, to dive, to fish, and to ensure the survival of newborns. Individuals do not live to become really old, because when they are no longer able to keep up with their group, they will be left behind, susceptible to predation. This possibility is a certainty for spinner dolphins, as discussed below.

It would seem even more difficult for zoologists to research the behavior of birds and bats, who fly freely in the air, than that of dolphins and whales, but such is not necessarily the case. Detailed research on nesting birds enables us to understand exactly how reproductive experience is acquired with age, because nests and nestlings are fixed during the reproductive period, allowing scientists to study these young and their parents in detail. In contrast, mammals move around at will, so it is usually impossible to know for sure how many young an older female has produced, who fathered them, and how well they survived.

Variability of Behavior within a Species

The behavior of older animals usually differs from that of prime adults, as the myriad examples in this and other chapters indicate. The males of most species, when they are older, are no longer able to dominate younger prime males by force. They lose fights, and then their high ranking in their troop's hierarchy; they become submissive subordinates, a distressing experience for them (see chapters 8 and 9). However, in a few species, the behavior of adults changes very little as they age. This chapter considers the aggression of older male elephants, and the sociality of spinner dolphins. It concludes with a contrasting look at the great variety of behaviors noted in human elders, behaviors that are shaped by their cultures.

MALE ELEPHANTS

The reason older male elephants remain pugnacious is that, unlike most adult mammals, they do not stop growing when they reach their prime, but continue to increase in size as they age. Because of his immense bulk and longer, heavier tusks, an older male elephant is likely to win a fight against a younger adult. Fully adult males periodically enter the stage of musth, when their blood has a high concentration of testosterone. They then move away from their group to hunt for a maternal herd containing a female in estrus with whom they can mate. At this time, two rivals will fight seriously, sometimes to the death. One such contest ranged over half a hectare as the two opponents, growling and screaming, pushed against each other and parried with their tusks, each aiming for a fatal thrust into the other's body (Buss 1990). If two elephants are shoving head to head, the one who loses his nerve and turns away gives his opponent the chance to kill him, usually by driving a tusk into his throat or flank. Joyce Poole (1996), who studied elephants for many years in Amboseli National Park, found that females in estrus much prefer to mate with the large bulls in musth rather than with young inexperienced males, so it is important for an older male to remain combative. Poole predicted that poaching and culling older males will jeopardize the reproductive rate of elephants, and thus the population's potential for recovery.

SPINNER DOLPHINS

Spinner dolphins (*Stenella longirostris*) are so intensely social that expert Kenneth Norris (1991) contended that a single animal could not be considered to be a whole individual. Indeed, if one of them were swimming alone in open water, he or she would soon be attacked and killed for food. In times of danger, all members of the group, old and young, form a closely packed school. Unlike orcas, whose families stay together for a lifetime, school memberships for spinner dolphins, drawn from a population of as many as 1000 individuals, are fluid, with some of them changing groups from day to day (Norris and Pryor 1991). These shifting associations are unusual, in that any one dolphin is often among animals not related to him or her.

The name "spinner" comes from an individual's habit of barking, and then leaping into the air and spinning around up to four times, similar to a skater's quad jump except that the dolphin hits the water on his back or side. Each lands with bubbles and a loud splash, which apparently help his schoolmates register his location on the surface of the ocean (Norris et al. 1991). Such jumps are not common when the animals are moving slowly and resting close together in shallow waters near land at midday, but become so when the group is feeding in open water at night, when members are spread out over a large area.

Most older animals of other species are somewhat distinctive from the rest of their group because of their age: they are larger/smaller and shrunken, more experienced, more wrinkly, or grayer than their peers. In contrast, older spinner dolphins are of interest because they look and act just like the younger adults. In an extensive study of the behavior of this Hawaiian species, Kenneth Norris and his colleagues photographed and sorted through 20,000 slides of local dolphins. By looking for distinctive individuals, they were able to identify several: Old Finger Dorsal (with a deformed dorsal fin) was recognized on 49 different occasions over a period of 11 years, and Old Four Nip was identified 69 times, usually baby-sitting the school's youngsters (Norris 1991). But we do not know how old these two distinctive dolphins were.

In his book *Dolphin Days* (1991), Kenneth Norris expounded on the importance of what he called the "magic envelope" to this rather dainty, slender-beaked, 65 kg (140 lb) species. These Hawaiian dolphins pass part of their day resting near shore, and then swim at dusk to open waters, where they spend most of their feeding time at constant risk from sharks and pygmy killer whales; many of the older spinners have extensive scars from unsuccessful shark attacks. Safety resides in their school's traveling formation, with all individuals swimming close together in the same direction—the "magic envelope" they form when predators are near. From the day they are born, all spinners live in a school of 60 or more. This reliance on their group means that if spinners are caught by fishermen, "they are docile, almost beyond belief, upon capture, and in captivity." They never fight against their captors, because this behavior is not in their psychic repertoire.

Inside the envelope are the dolphins themselves, as well as a cacophony of the often very loud sounds they make at or above our hearing range—screams, burst pulses (sometimes referred to as quacks, blats, banjo twangs, barks, or chuckles), nonharmonic and harmonic whistles whose meaning remains unclear, and echolocation clicks that enable them to sense where their peers, fish they might eat, and other objects are in the water around them (Brownlee and Norris 1994).

When animals travel in close, fast groups—as do these dolphins, shore birds such as dunlins, and many species of fish—their predators are unable to get a fix on any individual member, because they all look the same as they twist and turn, guided by some sensory integration system, and flash past too quickly. In tests done on flying dunlins, ornithologist W. K. Potts found that the maneuvering flock, changing directions constantly within seconds, was passing information among its members 2.6 times as fast as a predatory hawk could react (Norris 1991). The importance of schooling has been demonstrated by an experiment in which individual fish were dyed blue and then reintroduced into their school. These fish were attacked far more frequently by predators than were the other fish; the individuals near these blue fish were also more at risk, because they too had become a beacon on which the predators could focus (Norris 1991).

As this blue fish experiment showed, for schooling behavior to work effectively, the members must all look alike, so that a predator's eyes cannot become fixed on one individual. Both males and females must be about the same size (with spinner dolphins, the males and some females keep to the outside of the school to protect the youngsters, who remain in the center), and with the same color pattern. The spinner sexes are so similar that researchers have difficulty telling the males from the females in clear Hawaiian waters; they can do so only if they catch a glimpse of the genitalia or mammary slits from below. All the adults are slim, about 2 m (6 or 7 ft) long, with a black stripe down the back, gray stripes on the sides, and a white belly (Norris 1991). Slight differences visible in older males are the increased size of the hump behind the anus, and slightly taller, dark dorsal fins.

In contrast with these Hawaiian dolphins, who always live in schools, a different subspecies of spinner dolphins inhabits tuna grounds far from land (where thousands meet their death each year when they become entangled in the tuna nets). The water is murky there, and males of this dolphin subspecies have a large ventral hump and tall dorsal fins, which readily distinguish them from the females. These animals apparently do not need a school formation to protect them.

Since it is difficult to tell male spinners from females, and impossible to recognize older animals in the wild, almost nothing in particular is known about aged spinner dolphins. Elderly animals do what the rest of the adults do. Spinners often caress each other while swimming; Brownlee and Norris (1994) noted that "in active schools in daylight it was usual to see 30% or more of a school engaged in caressing patterns at any given time." This activity, which involves individuals rubbing their bodies together or one animal propelling another forward by inserting his or her beak into the other's genital slit (Johnson and Norris 1994), may continue into the night. Other activities include resting, swimming perhaps 65 km (40 mi) a day in search of fish, and males babysitting the young while females dive for food (Norris 1991). The oldest males are not necessarily the leaders of the school, because there seem to be no leaders, but they are probably among the males who herd the females and their young into safe shallow places to rest.

Norris (1991) imagined the end of an elderly spinner dolphin: "Mistaking her breathing for the first time in forty years, an old female inhales sea water instead of air, rolls onto her side and sinks into the water. Her companions, sensing that she is in trouble, swim to her and try, without success, to lift her with their beaks. Soon they sense she is dead and hurry forward to rejoin the school, the only safe haven they have ever known."

HUMANS

At the opposite end of the spectrum from animals in which older adults behave like younger ones is a species which has incredible variety both among its elders and in how these individuals are treated, since the latter affects how they behave. The species, of course, is *Homo sapiens*. In the book *Other Ways of Growing Old* (Amoss and Harrell 1981), anthropologists who have studied various cultures present information about how older people live in diverse societies. There is a marked difference between "primitive" societies and societies that are quickly becoming westernized.

For !Kung hunter-gatherers living in Botswana, elders are vital in five important areas (Biesele and Howell 1981). They are stewards of the rights to water and other resources in their area, they have the knowledge and skills that enable them and their families to live in desert areas, they are teachers and caretakers of children, they are healers who maintain the spiritual values of their people, and they have privileges in rites. The older men spend their days trapping, gathering foods, making crafts, visiting, and telling stories. The older women are also gatherers, craft-makers, minders of children, and often healers. Both men and women pass on their expertise to younger members in their daily activities, and are therefore indispensable members of their communities.

Among the Kirghiz of Afghanistan, older people have high self-esteem—it is a triumph for a person to have lived to old age in the country's harsh, high-altitude environment (Shahrani 1981). Elders among these pastoral nomads are important for their wisdom, which is thought to increase from year to year, thus bringing them increased respect and authority. They remain active in daily life even if they are physically less able, and become more religious as the time of their death approaches.

In some early societies, of course, there was little opportunity for older people to be useful. Among the Inuit, who relied entirely on hunting for their food, in times of famine elderly men and women sometimes left their family igloos to go out into the cold to die, so that the younger members of the group might survive.

The behavior of Micronesians inhabiting Etal Island in the South Pacific is in contrast to that of societies living as they have always lived (Nason 1981). Here, with increased Westernization, older people no longer have much self-esteem. With the introduction of democratic and Westernized political values, clan chiefs, usually elderly men, have lost their power. Western education has undermined the specialist roles and activities that used to make senior men and women respected, such as navigation, spirit mediumship and knot divining, magic, and warfare. In contrast to older people, who are being marginalized, educated young people who leave the island for a better future elsewhere are gaining respect for the wages they earn and send home to the island. With the latter's absence from Etal Island, the oldest people find it difficult to cope on their own.

Chipewyan [Chippewa or Ojibwa] First Nation Indians come from forebears who, for centuries, survived by hunting the vast herds of caribou who roamed the Canadian tundra. Now, many have been forced to settle down in small fixed communities, two of which, in northern Saskatchewan, have been studied by Henry Sharp (1981). He found that the Chippewa differentiated between "senior people" and "old people." The former were proud individuals distinguished by their family lineage or respected for their personal attainments. But old people were those who could no longer carry on as they had in the past, becoming a burden to the group, and often to themselves. Although men in their prime were dominant over their wives—given their physical strength, ability to obtain meat through hunting, and the backing of cultural traditions—over time their influence faded, while that of their wives increased because of the growth of their children and grandchildren.

Older Chippewa women retain their place in society much longer than do older men, even if their strength is failing. First, they often "adopt" grandchildren for different lengths of time, treating them as their own and being accepted as their mothers. These children can be useful in carrying wood and hauling water, reducing a woman's work load. Second, elder women retain their expertise in making handicrafts (largely beadwork and sewing), for which they have had a lifetime of practice. Even if they can no longer manage the intricacies themselves, they can still teach the younger women. Third, older women remain important because of their long experience in preparing food. All of these attributes keep elderly women in the mainstream of community life. However, men who can no longer hunt have largely lost their ability to command influence or respect in the group, because men's affairs are centered upon the bush rather than the social life of the village. For many men, who are often heavy smokers, old age begins as early as their late forties or early fifties, when they begin to suffer from lung or heart disease.

A wonderful advance for some older people is that many First Nations Indian elders in Canada are now being honored for their knowledge of their group's culture before it was heavily affected by European influence. They teach their language to young people, and have become vital as repositories of native lore. For example, in their past, Coast Salish Indian elders from Washington State and British Columbia were valued for their knowledge and expertise in such activities as producing food, building canoes, and making clothing. Few younger people are now interested in such things, but the elderly have gained prestige in a new way. Their claim to near-exclusive knowledge of the traditions of the past gives them great power (Amoss 1981). Their recall of native religious expression enables them to be the focus of Indian identity, and they are active in seeking to revive Indian culture among the young. The memories of older people, raised in oral traditions, are now often vital in working out treaties for aboriginal nations in Canada. If a First Nation can prove that their ancestors lived in specific areas of a province, this can affect treaties that may bring in many millions of dollars to a small community (Venne 1998). Far from being impotent victims of old age, many elders are becoming agents of positive activity that enhances society.

For all human societies, even those where nowadays elders are disparaged because of their age, seniors have been vital individuals who pass along their wisdom and advice to the next generations on such matters as child care, health care, family problems, matchmaking, religious rites, and social affairs (Waterhouse 1983).

IN SUMMARY, ANIMALS LIVING in social groups are more successful than they would be living on their own. What we know about each species is greatly affected by the environment each inhabits: on land which we share with them, in the air where they are more difficult to observe, and in water where daily exchanges are largely unknown. In a few species, such as elephants (male) and dolphins, interactions of individuals within their groups apparently change little as they age, while in human beings, by comparison, culture hugely affects how old people are perceived and how they behave. The next chapter addresses how essential the experience of female matriarch elephants is to their herd.

CHAPTER 3

The Wisdom of Elders

A YOUNG "ROGUE" ELEPHANT ASSAULTS tourists in the Pilanesberg Game Reserve, South Africa, in 1996. The next day he kills the professional hunter sent out to shoot him. Out-of-control young males attack, try to rape, and gore to death nineteen white rhinos (L. Watson 2002). What is going on here?

Lack of Supervision

In the 1980s, officials in Kruger National Park, South Africa, shot a number of elephants because they were becoming too numerous for their habitat. Men aimed darts at adults from the air, then gunned them down as they lay anesthetized on the ground. The youngsters who watched this nightmare were then rounded up and sent to parks and reserves that lacked elephants (L. Watson 2002).

Problems such as those described at the head of this chapter erupted 15 years later, when the young males from this group were about 20 and beginning to produce serious amounts of the hormone testosterone. Their pathological behavior was related in part to the earlier killing of their family members and the stress of translocation they had suffered; scientists equated it to human PTSD (post-traumatic stress disorder), which involves "abnormal startle response, depression, unpredictable asocial behavior and hyperaggression" (Bradshaw et al. 2005; Meredith 2007). However, the bizarre aggression of the young males mainly reflected a lack of adult supervision when they were growing up as orphans. Under normal conditions, when the males were juveniles, they would have been socialized by matriarchs. At puberty, when young teenage bulls left the herd into which they were born, most would have gathered together casually and hung out with a variety of other males for the rest of their lives. During this time, the young are guided in their behavior to some extent by older males. Aging bulls are important in quieting young males during their first period of musth, when they develop an unfocussed aggression triggered by increased testosterone. These seniors spar with them, dissipating some of their hostility and forcing them to calm down. Whereas in most species older males do not compete against younger prime males, this is not the case for elephants, where adults and their tusks continue to grow with age. Even an elderly male can be a formidable opponent, and it is these dominant males who usually sire the calves.

This chapter about elephant behavior is entitled Wisdom, because it depicts how wise older elephants are, due to experiences garnered over a long life. Like male elephants, females also continue to grow after they reach adulthood. The largest females, the ones with the most experience, are the most dominant (Archie et al. 2006). Their reign as a group leader is not usurped by younger animals—those with less experience, but physically bigger and stronger—as it is in most social species. Elephant matriarchs share their vast knowledge with younger individuals, and they are responsible for maintaining order and harmony within their herd. If a young calf gets out of line, a matriarch will give him a disciplinary whack with her trunk; if there is a dispute, she quickly quells it.

When older animals are targeted preferentially to be slaughtered in a government cull or by poaching, the results for the family are devastating, rendering the herds dysfunctional. They have lost the cultural, learned information of the matriarchs: how to engage socially with other elephants, how to resolve conflict, how to communicate effectively, and, most importantly, how to physically survive in the wild. The depleted elephant group of youngsters and subadults will now flee in terror from human beings. They have reason to panic, because without guidance from older matriarchs, mayhem results. With its elephant mother gone, statistics show that every infant under the age of 2 will die, those orphaned between the ages of 3 and 5 have a 30 percent chance of survival, and only half of those orphaned between the ages of 6 and 10 will survive (Chadwick 1992). Thus a lack of maternal supervision can devastate the young in a herd. In no other species is the wisdom of elders, gained in earlier experiences, so vital.

Importance of Matriarchs

The behavior of older female African elephants (*Loxodonta africana*) is admirable. The matriarch makes all the daily decisions for her herd, a group of females and their young so intensely social that the females spend their entire lives together. A matriarch is often a great-great-grandmother, with a calf at foot, along with several older young. Every member of the herd respects and obeys her as she oversees the elephant activity taking place around her. They will never be out of sight or earshot of their matriarch, their mothers, their own young, their aunts, their siblings, their nieces, and their young nephews—a nightmare concept to those of us who value our independence.

Longevity is important; it ensures that the older, biggest elephants with the most experience and social knowledge are the best leaders (female) and fighters (male). The average age of a number of matriarchs shot in one cull was about 49 (Laws et al. 1975), and although old elephants have been defined as those ranging from 40 to 55 years of age (Gröning and Saller 1999), some matriarchs are over 60 years old. These latter two researchers considered elephants over 55 to be "senile," but this is an unfortunate translation from the original German text. Elephants can live into their seventies, and there is no reason to assume that these seniors are somewhat incompetent, as the word "senile" implies. Sylvia Sikes (1971) reported that one impressive old female—a matriarch of 60 or even 70—had three calves at foot, the youngest still sucking. She was in absolute and aggressive control of her herd (or was until Sikes shot her), despite an autopsy which showed that she suffered from varicose veins in her legs and atheroma of the aorta and coronary arteries.

Cynthia Moss (1988), in her wonderful study of wild elephants over many years, was able to individually recognize 650 elephants in Kenya's Amboseli National Park and work out their various relationships. For a large herd, there may be two matriarchs, often sisters, as leaders, with both of them respected and obeyed by all herd members. A herd may also include a former matriarch, now too decrepit, shrunken, and slow to do more than keep up with the group most of the time. Where there is plenty of food, several families making up a small herd of 20 or so may join with other, probably related families, at times forming groups of a hundred or more. The more animals, the more likely the group is to detect danger and take measures against it. When there is a threat to a composite herd, different sets of elephants immediately rush to bunch around their particular matriarch and keep their own young well protected, thereby revealing the internal composition of such large groupings (Laws et al. 1975). If there are large males among the threatened group, these show their independence by rushing off in different directions without regard to their fellows.

Katy Payne (1998) discovered that as the matriarch rounds up her herd to start out on the day's foraging, she gives a distinctive "let's go" rumble, too low to be heard by human ears but picked up by a microphone. When a herd of elephants lacks a matriarch, they vacillate about which direction to take; the microphones indicate that no individual gives the "let's go" rumble, so there are a number of false starts before the elephants agree on a route. Being highly social animals, it is not an option for individuals to go off in different directions. Elephants communicate not only by low-frequency calls transmitted through the air, but also through sounds propagated in the ground (O'Connell-Rodwell et al. 2004). Such seismically-transmitted warning calls may travel farther than their airborne counterparts. Female African elephants can recognize a contact call as belonging to a family or bond group member over distances as great as 2.5 km (1.5 mi).

RAISING YOUNG ELEPHANTS

If it takes a village to raise a child, it certainly takes an elephant herd to raise a baby elephant. Elephants are civic minded, with each mother looking out for other calves as well as her own. Each mother supplies her own young with milk, but additional adult relatives are instrumental in helping look after him or her. Many infants spend most of their time with an aunt or grandmother who guides them, providing them with information based on her experiences and knowledge of the environment—what behavior is acceptable, what plants are edible, what constitutes danger. Sometimes her body provides shade for a hot youngster (K. Payne 1998). One young male whose behavior Cynthia Moss (1988) observed in Amboseli National Park spent far more time with his grandmother, Teresia, the 58-year-old female who had previously been the herd matriarch, than he did with his own mother Theodora. He would suckle from Theodora, then go to Teresia each day to hang out. Teresia also remained in close touch with her much larger son, Tolstoy, who showed no signs of wanting to leave the herd, even though he was 13 years old.

Teresia suffered a terrible end to her life (Moss 1998). When she was 62, three young Masai men chased her herd, which had wandered outside Amboseli Park. Because she was the slowest runner, they threw their three spears into her neck and shoulders. When she charged them, they ran away in panic. Teresia was too badly injured to return to the park and her family. Instead, she hid among vegetation, her wounds festering and painful. Eventually she crumpled to the ground and died. Moss was devastated by her death, but still thought it preferable to Teresia being shot during a cull, when her whole family would have been wiped out: "all her female descendants, most of her genes, nearly everything she had lived sixty-two years for."

Jezebel, about 60 years old, was another matriarch of Amboseli National Park. She had a 7-year-old son, probably her last offspring because of her age, who on occasion still nursed (Chadwick 1992). Sometimes Jezebel oversaw a game of catch between her 10-year-old nephew, Joshua, and researcher Joyce Poole (1996). When Poole threw a piece of buffalo dung or her sandal toward him, Joshua picked it up with his trunk and chucked it back.

A female in her prime produces a baby about every four years. Matriarchs may produce calves regularly, just as their younger sisters do, but after about age 40 the infants are spaced further apart. In every case, aides are vital in helping a mother train an infant, her 6- or 7-year-old, and her teenager in puberty. The mother could not do this all by herself, especially if she is a matriarch charged with the health and safety of the entire herd. When a large herd is on the move, the leading matriarch takes care to proceed at a pace slow enough for the youngest member; the matriarch bringing up the rear is ready to help any infant who falls down or has trouble clambering up a hill or crossing a river (Gröning and Saller 1999). Calves learn a great deal directly from adults, and about their environment, as they grow—their brain at birth is only 35 percent of its eventual adult weight (in newborn human babies it is 23 percent) (Chadwick 1992).[1]

One of a matriarch's many responsibilities is to rescue calves in distress. When a female elephant in Uganda was dying from gunshot wounds, the rest of the herd gathered protectively around her, all of them agitated, especially her young son (Buss 1990). When she died, her traumatized infant remained pressed against her body, even after the matriarch had partially covered the head and shoulders with grass and other vegetation, as if in symbolic burial. The rest of the herd then slowly departed, but when it was some distance away the matriarch, probably the dead cow's mother, returned and used her trunk to gently pry the youngster away from the corpse. Twice the distraught infant escaped from her to go back to his dead mother, but finally she was successful in steering her grandson, safely protected between her forelegs, to the waiting herd members.

CARE OF THE HERD

In addition to calves, matriarchs also look out for wounded or ill members of their herd. When Irven Buss (1990) darted or shot individuals who then collapsed, the herd matriarch would often scream and race to the fallen animal, trying to raise him or her. Then she would trumpet and charge furiously in various directions, anxious to attack whatever had caused this catastrophe.

A matriarch's concern for herd members is portrayed in Karl Gröning and Martin Saller's elephantine (3.6 kg or 8 lb) book *Elephants: A Cultural and Natural History* (1999). In six color photographs, they show (1) an old female separating from the herd, (2) the others stopping to wait for her, (3) the matriarch coming back to her, (4) the female collapsing, perhaps from a heart attack, (5) her 15 or so friends clustering around her as if in sorrow, and then (6) the herd moving on because there is nothing else to be done.

Herd matriarchs are always alert to possible danger. In Tanganyika (now Tanzania), one man watched a herd of elephants, including two infants, crowd into a watering hole to drink and bathe. Suddenly the matriarch reached into the muddy water, grabbed a 4.5 m (15 ft) crocodile with her trunk, and carried it to the beach. She held it high over her head, a weight of well over 450 kg (1000 lb), before smashing it again and again on the ground. Then she moved farther inland to a tree, where she flailed the crocodile against the trunk. Finally she and another cow stomped on it as it lay on the ground. The enemy was unrecognizable when they were finished with it (Buss 1990).

However, most attacks come from hunters, against whom elephants are nearly helpless. If poachers are active in an area, the matriarch will lead the herd away at the first scent of human beings. If the herd is attacked, she will often rush forward to defend the others, and so be one of the first to be gunned down (Chadwick 1992).

In 1950, a man hired to shoot elephants killed three of them, and wounded an immense bull who had been standing near an elephant herd. Its elderly matriarch, with her gaunt, sunken features, led her herd away from danger. However, five elephants lagged behind, going at half her pace. When the hired gun studied this slow group with his binoculars, he saw that it was composed of two large females on either side of the very large old wounded male, leaning inward to support him with their shoulders. Behind them were two more females, with their foreheads pushing against his rump (Buss 1990). (Given the less social nature of large males, they would never take part in such rescues.)

This behavior is in marked contrast with that of a nineteenth-century "sportsman," R. Gordon Cummings, when he encountered a wounded male—well, actually the male was injured because Cummings had just shot him in the shoulder, so he could not escape. First, Cummings made himself some coffee while the crippled elephant leaned against a tree, helpless and in pain. Then he decided to do a torture experiment, to see which were an elephant's most vulnerable areas. He approached the animal and shot him in different parts of the head. The elephant remained still, except for large tears that trickled from his eyes, and his trunk tip, which probed the various bullet holes. Then Cummings, perhaps at last rather embarrassed by his own behavior, shot the animal nine times behind the shoulder, and he finally fell over dead (cited in Masson and McCarthy 1995).

The picture is not always sweetness and light in elephant society. During Zimbabwe's terrible drought in 1981, water was at a premium. Elephants had to feed for at least 16 hours a day to survive, and then walk the long distance to a water hole that might be dry. Sometimes a matriarch had to tusk out a hole in sand, so water could slowly seep in (Buss 1990). Fights broke out over the water; mothers did not push their own calves out of the way, but they had little time for other weak animals. A warden reported that "bulls shoved little orphans aside, and the matriarchs and other older females tusked the hell out of subordinates." Many animals died of thirst (Chadwick 1992). However, when a calf died, its mother sometimes refused to leave its body, and so died herself from lack of water (Buss 1990).

Advantages of Older Matriarchs

Herds led by the oldest matriarchs in Amboseli National Park have the best reproductive success, with more infants produced and raised there than in herds led by younger females. In experiments extending over seven years, Karen McComb and her colleagues (2001) found that elderly matriarchs were able to distinguish a large number of other elephants (about 100 others in the park) by their contact calls. Younger matriarchs were less proficient in recognizing the calls of individuals. Having this social knowledge meant that at the approach of other elephants, the older matriarchs were more efficient in deciding when defensive activity, such as moving the young into clusters, was appropriate, and when cooperation was the best response. The researchers wrote that "families with old matriarchs benefit reproductively because their matriarchs target caution at the appropriate individuals: callers that are strange to them." They noted that previous researchers had speculated that elephants derived their fitness benefits from ecological knowledge, attained from experiences during their lifetimes, and concluded that "aging may also influence reproductive success through its effects on the acquisition of social knowledge and that the possession of enhanced discriminatory abilities by the oldest individual in a group of advanced social mammals can influence the social knowledge of the group as a whole." Because of this, in "harvesting" campaigns to reduce the number of elephants in places where they are destroying the environment, killing the older matriarchs makes the herd socially dysfunctional.

A MATRIARCHY, WHICH MAKES for gregariousness and family cohesion, relieves the male of parental responsibility. As in elephant herds, it often features large groups, with unselfishness and care for the young as its imperatives. In contrast, patriarchal groups, such as gorilla troops, are never large. The dominant male, who is sexually jealous, will not allow other adult males within his group, and can therefore only retain a limited number of females. Fraser Darling (1969) contended "that the matriarchal system in animal life, being selfless, is a move towards the development of an ethical system." It valorizes sociality, and certainly works wonderfully well for elephants, as well as for hundreds of other species.

CHAPTER 4

Leaders

C HAPTER 3 DISCUSSED HOW OLDER ELEPHANTS make superb leaders, in large part because of their age and experience. Various whale leaders are undoubtedly as experienced and wise as elephant matriarchs, but we must infer this, because we are not able to see much of what they do in the dark waters where they live. This chapter valorizes older and experienced female orcas and sperm whales, two large and long-lived cetaceans about whom more is known than most aquatic mammals. Elderly females of these two species, like older female elephants, are born into intensely social groups so all-encompassing that the members will never be on their own until the day they die. However, orca males and females live together; in sperm whales, they lead largely separate lives. The less-studied baleen right whale is also considered, because aged individuals seem, by inference, to be leaders, although this has not been verified. The leadership or past leadership of some alpha animals will be mentioned in future chapters, but this chapter ends with varied information on leaders of terrestrial species: sheep, goats, mountain goats, red deer, elk, baboons, and langurs.

Aquatic Leaders

ORCAS

Tourists leaning over the rails of ferries crisscrossing the coastal waters of southwest British Columbia are thrilled to see a pod of orcas (*Orcinus orca*) sail past. (I refer to this species as "orca" rather than "killer whale"; the word "killer" has a negative connotation which could, in reality, be applied to all predator species, including human beings.) Not that the ferry passengers usually see much—only dorsal fins cutting through the waves: those of the large males a rigid 1.5 m (5 ft), those of the smaller adult females 0.6 m (2 ft) shorter. If there were a whale researcher on board, he or she could tell the passengers that all the hundreds of orcas living in the area are known individually to scientists, each recognizable by characteristics such as the shape of the dorsal fin, nicks or scratches on this fin, and saddle markings (Hoyt 1990).

Scientists may know which orcas swim about together, but otherwise their work is largely done while these researchers are blindfolded, in effect, except for the occasional glimpse of a fin or a fluke. Nor can they decipher what the orcas are saying to each other by listening to their complicated whistles, squawks, squeals, and clicks through hydrophones, although scientists now often know who the vocalist is by his or her signature call.

To calculate the approximate age of an orca female in the Johnstone Strait area of British Columbia, researchers counted the number of young who accompanied her on her ceaseless roaming through its many passageways. A female orca becomes sexually mature at the age of 12, a gestation period is 16 months, her first young is usually lost, and she nurses each newborn for several years before pregnancy and the birth of her next offspring (Morton 2002). A female can live to be 60 or more in the wild if she is not killed by human beings, her only predator. However, one expert described a captured female orca as "old" even though she was only in her thirties, and even at age 25 an orca's teeth can be extremely worn (Hoyt 1990).

Males are much more difficult to age. At 20, their enormous dorsal fins are fully grown, so at least a minimum age can be estimated if there are photos taken in earlier years of individuals with adult fins. Perhaps in part because these protuberances cause a terrific drag for the males as they swim and dive, using up much energy, they die far sooner than the females, with a lifespan of only about 40 years (Morton 2002).

The much-studied orcas resident in British Columbia and Washington State waters live in matriarchies, the members staying together during the lifetime of their elderly leader, with neither males nor females transferring to other pods. The lifelong bond of a mother and her offspring form the basis of whale social organization. Adults of pods meet those of other pods now and then, so there is no problem of inbreeding. However, so-called transient orcas, which have a wider distribution and who feed largely on seals and porpoises rather than fish, do not have such a stable system. Because their source of food is more irregular, the young leave their natal pod when they mature, and families dissolve with each generation (Knudtson 1996).

Not long before Alexandra Morton (2002) moved to the coastal waters of British Columbia to study orcas in 1979, these animals were being shot for sport. Children were encouraged to throw rocks at them. The government tried to cull their numbers using machine guns. Fortunately, their status has improved drastically since then, but, shamefully, orcas are still kept in aquariums, literal prisons for animals who committed no crime. At least many people now admire them, and are aware of their intelligence and complex social relationships.

Older females likely have stopped reproducing, but they remain invaluable to the pod (Knudtson 1996). As leaders, they not only guide the way to traditional sites such as rubbing areas, resting places, or salmon runs, but they also pass on knowledge they garnered over a lifetime— hunting strategies, echolocation skills, and the vocal dialects of nearby whale groups. They seem to teach these things by demonstration, since orca whales have an amazing ability to mimic, evident in aquarium or marine park displays. Aging females also help care for pod young while their mothers are feeding or resting.

Older individuals, especially females, sometimes travel through their range with a younger relative, apparently to show them special areas to remember among the maze of islands and waterways, many hundreds of kilometers long, where they live (Peterson 1998). A female over 70 toured the area with her son one recent summer. As they swam along, every nearby whale came up to the matriarch to spend a few minutes with her before moving back to allow her to swim on. It was as if she were making an old-fashioned progress through the waters where she had lived for her entire long life.

Morton, who studied orcas in this area for over 20 years, described one female leader in her book *Listening to Whales* (2002). Eve,[1] who was about 42 years old when Morton first saw her, was considered old because she had stopped reproducing nine years earlier. Eve belonged to the same pod, A5, as the young female Corky, the same captive Corky whom Morton had known earlier from working at Marineland near Los Angeles. Ten years before, 12 whales from Eve's pod had been herded into Pender Harbour, north of Vancouver, by boats; 6, including Corky, had been captured and shipped to various marine parks and aquariums, wiping out an entire generation of the pod's young animals in the wild. Most of these young captive whales soon died in their small pool prisons.

Eve was doubtless traumatized by this event, growing up at a time when motorized boats buzzing around whale families sometimes shot at them, and by an early encounter with a boat. She had deep hollows carved out of her back where the propeller had nearly killed her, its blade slicing down almost to her spine. Eve was still wary of boats, keeping her distance from them. She often foraged far ahead of her group, perhaps because she had become intolerant of the rambunctious and noisy younger pod members.

Morton, in turn, would be traumatized by an event involving Eve. Whales and dolphins have a reputation for rescuing people in trouble in the water by pushing them up toward the air or in toward shore—although there is no way of knowing if people are sometimes pushed down and away from shore—but this is not what happened to her husband (Morton 2002). In September 1986, Alexandra and Robin were filming whales near Robson Bight in Johnstone Strait, a beautiful area where kayakers such as myself have stopped to rest. Robin was stationed in diving gear 9 m (30 ft) offshore and 9 m underwater, while Alexandra and her young son idled nearby in their zodiac. Suddenly, without warning, Eve surfaced nearby without the rest of her pod in tow, something that was highly unusual. She dove toward Robin, whose camera was set up on the ocean floor, and then resurfaced again and charged toward the zodiac before turning away. Alexandra was amazed that she had surfaced twice in such a short time, another unusual occurrence.

Alexandra waited for her husband to emerge from the water in the next few minutes, excited to ask if he had shot any footage of Eve, but he never came up again. She found his body underwater, lying prostrate beside his camera. At first, it was thought that he had somehow been killed by the whale, but later the police found a malfunction in his diving equipment, which deprived him of oxygen. It may be that although Eve did not push the dying (or already dead) Robin toward air, she was attempting to alert Alexandra that something was wrong. It was not direct aid, but it could have been an indirect means of trying to help.

In 1990, Eve died of unknown causes at age 53, her body cast up on a lonely beach. Her adult sons Top Notch and Foster swam around and around in the area, calling for her, but for the first time in their lives she did not answer. Alexandra helped tie the corpse by the tail, so it could be towed through the water to Telegraph Cove. There, biologists autopsied Eve's body and preserved her skeleton. Most orca whales die without their carcasses ever being found. Eve's was able to provide information new to science, because her stomach contained the remnants of 59 individual fish, many of them bottom feeders, belonging to 13 different species (Morton 2002). The anatomy of her blowhole was analyzed, and her pectoral fins were dissected to reveal the hidden small fingers within them. Her skeleton was reconstructed, so it could be put on public display in the Sidney Museum on Vancouver Island.

Another notable older female was Stubbs, the leader of a pod of 16 whales described by Erich Hoyt in his book *The Whale Called Killer* (1990). Stubbs was instantly recognizable because her dorsal fin had been hacked off, leaving mangled edges in its place, probably the result of a collision with a boat propeller. She wasn't especially wary of boats, although she tended to keep more to herself than the other whales did. She swam slowly, never speeding or leaping upward, spending long periods lying on the surface of the water, perhaps in a kelp bed, while the rest of the pod swam on ahead. When the other whales slowed down, she caught up with them. She took more breaths than the others and sometimes coasted on the surface, belly side up, letting the tide carry her along. She was gentle; when there was a canoe in the area, she took care not to upset it when she came up from a dive, as if she knew it to be a fragile craft.

Besides leadership, older orcas have another role in their pod, namely babysitting the young each day while their mothers dive for fish. These clusters of youngsters, along with an aging bull or other elders, may actually serve as a form of school, where the young are taught such things as how to catch salmon. Anyone who has studied the behavior of orcas in depth no longer thinks of them as killers. Rather, they are seen as the sociable, intelligent, and efficient animals that they are.

SPERM WHALES

The sperm whale (*Physeter macrocephalus*) is an oddity to human eyes, with its (comparatively) tiny lower jaw, and its monstrous head, largely composed of spermaceti oil, so named because it is the color of semen. The spermaceti organ, which comprises over a third of its body, is believed to be involved in the production of echolocation clicks, which enable an individual to "see" and catch fish and squid in the black ocean depths, often as much as 2500 m (8200 ft) below the surface (Gordon 1998). Their vocalizations are incredibly loud, as befits the huge structure that produces them.

At 18 m (60 ft) in length, sperm whales resemble large baleen whales, but they are actually by far the largest member of the dolphin group, with lower and (less often) upper teeth, rather than baleen plates, in their mouths to obtain food. (Odontocetes are usually aged by counting the layers in their teeth, but this is inexact in older sperm whales, where the pulp cavity of their teeth has closed.) Researchers who are able to identify thousands of individuals by the indentations on the edge of their flukes note that the older the whales are, the more little patches of rough skin—calluses—they have on their small dorsal fins (Gordon 1998).

In his book *Moby Dick*, Herman Melville portrayed the sperm whale as the terror of the seas. The popular *Larousse Encyclopedia of Animal Life* (1967) described them as aggressive animals that will "assault a whole boat manned by whalers, crushing it like matchwood in their powerful jaws." In reality, unless they are being harpooned and harassed by human hunters, sperm whales are by nature timid and inoffensive (Whitehead 2003). One group of about 20 resting whales all dove together in a defensive maneuver when a fur seal suddenly surfaced near them; however, they may have learned such cautious behavior from having relatives massacred by human whalers before 1986 (or more recently illegally). Most would not have seen the slaughter taking place, but many would know about it from their ability to communicate through water over huge distances with their low calls.

Because of their long experience, older animals serve as their group's leaders. Elderly sperm whale females, for example, live in social nomadic maternal pods which include females of all ages and young males not yet ready to leave their mothers, just as elephant herds do. These individuals always hang out together, swimming, resting, feeding, vocalizing, leaping into the air, and sometimes rubbing against each other, loosening skin which sloughs off into the water—skin patches currently very useful if collected for DNA testing (Gordon 1998). The elder females are integral members of their pod, but they rarely have a calf to sustain. Females reproduce less often as they age—an average of once every 5 years in their prime, and rarely, perhaps once every 15 years, when they are in their forties.

Since females can live into their eighties, but usually only reproduce until their forties, they have an ample number of years to devote to their relatives (Whitehead 2003). Older females, sometimes called "sages," are vital to a pod, because they remember the location of good "feeding grounds," a term which sounds solid, but in fact describes biologically productive areas a few hundred to a few thousand kilometers across that are scattered throughout tropical and temperate waters—not just any waters, but those at least a kilometer (0.6 mi) deep. Sperm whale pods are always on the move, traveling about 35,000 km (22,000 mi) a year, as elders lead them from one good foraging area to another as the available resources of each feeding ground change.

These sages pass along to the young cultural information that they have collected over the years (Whitehead 2003). This includes suitable movement patterns, what creatures make good eating, and vocal dialects, which may convey valuable information—such as bypassing possible reproduction partners who have the same dialect, to avoid inbreeding.

Sperm whales have a social structure similar to that of elephants, with small groups of females caring for and defending their young together, traveling long distances in search of food, and randomly encountering other sperm whales. Recent studies of elephants reveal the importance of older matriarchs to their herds, and the chaos that exists when they have been killed (see chapter 3). This same dysfunction may have occurred in whale pods. Sperm whales off the coasts of Peru, Chile, Japan, and even northwestern Europe all have lower birth rates than would be expected in normal pods, such as in the Caribbean, where whales were not killed. This may be because of the slaughter of larger, older individuals, practiced by whalers until 1986. The sperm whale groups may have "lost their social knowledge and may be less successful" for this reason (Pennisi 2001).

RIGHT WHALES

Could old (but not the oldest) right whales (*Eubalaena glacialis*), with their vast experience of traveling the world's oceans, be leaders? We are not sure, but it seems so. Why else would oldsters who are no longer breeding make the long, exhausting yearly migrations to warm water from the polar regions where they mainly feed? In his book *Among Whales* (1995), Roger Payne (the former husband of Katy Payne—she later turned from working on the bioacoustics of whales to studying the vocalizations of elephants) described his work among baleen right whales off the coast of Argentina, where they give birth.

Older whales are those who best remember the location of choice food sources, and can lead other members of their group to them (Clapham 2004), although no one yet understands how they know where food is, given that the dense masses of plankton they need to survive shift about in the vast expanse of the oceans. Do they have chemoreceptors that can sense the presence of plankton? Or is it their sense of smell? (Baleen whales have a relatively larger olfactory lobe in their brain than do toothed whales and dolphins, who both find food by echolocation. Even to human beings, a large patch of plankton has a distinct smell.) Or do the elders know that food is concentrated at certain temperature gradients, which they can follow in their journeys?

If older whales aren't leaders, why do they migrate, since this is not necessary if they are no longer reproducing? Whaling records show that older, and very old, individuals were present on breeding grounds. Payne (1995) himself watched elderly female southern right whales, with their wrinkled skin, from a cliff overlooking the Atlantic Ocean—huge animals (bigger than the males) who no longer had calves, moved slowly, and interacted little with other whales in the bay. As far as the Paynes could tell, the presence of oldsters in breeding areas did not directly benefit younger generations of whales.

Perhaps older right whales migrate from force of habit, because it is something they have done each year during their long lives. Or perhaps these whales migrate because there is always some food to be found outside of the polar regions, which will shorten the period of fasting before they again head south (or north for northern right whales) to feast on krill. Maybe migration, which involves long-distance swimming, helps slough off the barnacles attached to their skin. Barnacles are a menace, because their rough shape affects the streamlined contours of a whale and slows down its speed. However, at least in humpback whales, barnacles are beneficial; their sharp edges can tear an opponent's skin in a fight (R. Payne 1995).

Why do right whales migrate at all? Payne did not believe that their young needed to be born in warm water, because they were not yet equipped to handle the cold. Rather, he thought they needed calm waters—to enable them to learn to breath properly, without inspiring water during crashing storms—and shallow depths, for protection against predators such as sharks and orcas. There is obviously much more we need to know about whale behavior.

Terrestrial Leaders

Most social animals have leaders who decide where their group will go to feed, drink, and sleep. Older animals make good leaders, as they have the most experience with both the environment and their group's capacities. With good leadership, a troop or herd will thrive; with poor leadership, it likely will not. There is a popular belief, undoubtedly based on human culture, that leaders gain their status by virtue of competition and fighting. This is true for males in almost all of the species where males and females remain together throughout the year, including lions, wolves, common baboons, and chimpanzees (but not hyena clans, where the females are dominant). Alpha animals are at the peak of their strength, and do not include older individuals, so these leaders are not of interest here. However, not all group leaders, especially those in charge of females and their young, are in their physical prime.

In his influential study of sheep, John Paul Scott (1945) argued that leadership can be gained not by dominance and aggression, but through social behavior, which depends to a large extent on training. Over a four-year period, he observed the behavior of a small flock of domestic sheep (*Ovis aries*) living in a 2-hectare (5-acre) pasture, free of human interference. He noticed that newborn lambs followed their mothers, who rewarded them with milk. This following behavior was not largely instinctual, but instead was a product of the lamb's training. Orphan lambs were not trained to follow, and so were much more independent. While lambs and older youngsters followed their mothers, as they had been trained to do, rams followed the ewes for another reason, hoping for sex. Consequently, "leadership of the flock went to an elderly ewe, inferior in strength and fighting ability to almost any ram, and often inferior to the younger ewes. This position was achieved mainly by the care and feeding of her descendants without, as far as the observer can tell, any instance of violence toward her offspring."

Elma Williams, in her book *Valley of Animals* (1963), reported that her free-range goats (*Capra hircus*) are led by an older matriarch, too. Thirteen-year-old Niall was the undisputed leader of her 12-member herd. At day's end, after browsing and grazing with the herd in a field, Niall would suddenly raise herself to her full height, point her beard to the sky, and then lower it to the ground. Then she would move slowly forward, saying "urh-urh" or "come on," which had the immediate effect of rousing the other goats to drift after her and then follow in single file, keeping to her pace if she paused to nibble at a bush. Williams called this line the "Goat Train." If there was a storm, the goat train became the goat express, with all the animals, still in single file, rushing forward after Niall to reach shelter. If individual goats, absorbed in eating, managed to miss the train, they quickly became panic stricken and rushed about, giving plaintive bleats as they tried to catch up.

In a female herd of wild mountain goats (*Oreamnos americanus*), an older, experienced female will also lead, knowing how best to circumvent dangerous landslide or avalanche areas during spring and fall migrations, and where choice vegetation grows. Adult females of this species have a linear and stable dominance hierarchy, with dominant individuals having increased kid production (Cote and Festa-Bianchet 2001). The guidance of the leader will help keep her younger sisters, daughters, nieces, and grandchildren safe. "If the tendency to obey a leader is transmitted in the genes, the wise doe's kindred will pass the tendency down through the generations" (Thomas 1994).

The leader of a matriarchal herd of red deer hinds (*Cervus elaphus*) and their young is either a mature or an elderly female. Other females do not challenge her leadership, but if she stops having a yearly calf, she soon gives up this responsibility. Otherwise she remains leader until she dies. Fraser Darling (1969), who studied these deer in Scotland, suggested that as a regular breeder, part of the leader's maternal emotions will have embraced the herd as well as her own young. Related to this, Scott (1945) suggested that older domestic sheep, with their numerous descendants, become leaders partly because their own tendencies to follow are reduced while they are caring for their young.

The well-known naturalist Andy Russell (1977) described the leadership of Old Buck, a cow elk (the same species as the European red deer) in the Rocky Mountains, who was head of close to 100 females with their calves as well as a few yearling bulls. It was she who broke trail through deep snow, chose the route they would follow, and made quick decisions when danger threatened. She had won her authority not by fighting (although she used her flashing front feet to punish any cow whom she felt was out of line), but because she was strong and had long experience.

On the surface, older baboon males may not seem to be leaders, but in fact they usually are, as Hans Kummer (1995) discovered in his extensive field observations of hamadryas baboons (*Papio cynocephalus hamadryas*) in Ethiopia. When a troop is ready to set out on its daily travels, a young dynamic male marches off in one direction while an older male in the rear, who also has a specific direction in mind, may or may not follow, along with the rest of the group. If this elder male remains seated, the "leader" tries another direction and again checks to see if this one is what the other fellow has in mind. The two never openly disagree with each other, just exchange glances. Even if the young male has many females and young and the scrawny older male has none, the latter continues to direct movements from the rear of the baboon column. The teamwork finally breaks down when the oldster is too weak to keep up with the others. Undoubtedly the young male learns from the older male's decisions which will be based on the variety of food, water, and sleeping sources available, and on the necessity of changing routes so that a leopard cannot hide along one of their predictable paths and ambush them. Rarely, an older male, such as Admiral, would not wait for all this shilly-shallying. By sweeping directly out of the center of the group with conspicuously long strides and not looking behind him, he could mobilize everyone to be underway in 10 seconds.

Langur monkey females lose dominance as they age, so that older animals are no longer central to the life of their troop; they tarry at the edge of the group, where they will not attract undue attention to themselves. However, elderly females continue to be respected for their far-reaching experience. They are the most likely members to know which trees in the area are fruiting at any one time, what gardens have fierce watchmen, where water is available, and which urban areas are most anti-monkey. Sarah Hrdy (1977), in her research in India, found that in almost every instance when a female decided in which direction a troop would travel, it was an older rather than a younger one, despite the former's low dominance ranking, which made her presence largely shunned in the core of the troop.

IN CONCLUSION, LEADERS ARE OFTEN older animals (male or female, depending on the species' social organization), who lead either directly or indirectly. Their one common feature is experience. They have all lived a long time, and their extensive experience—both in their environments and within their groups—helps them and their younger groupmates survive.

CHAPTER 5

Teaching and Learning

Teaching

ARE NONHUMAN ANIMALS TEACHERS? Is it possible to pinpoint activity that can be labeled "teaching" per se? In their book *How Monkeys See the World* (1990), Dorothy Cheney and Robert Seyfarth argued that monkeys (who are presumably smarter than most mammals) could acquire novel skills from others through observation, social enhancement, and trial-and-error learning, but doubted that they were able to imitate the behavior of others, and claimed that they were unable to teach. More recently, Bennett Galef and his colleagues (2005) invented a method that shows that a mother rat (*Rattus norvegicus*) does not teach her young what to eat. Other researchers, however, think that teaching does occur in nonhuman animals, although little is documented on the subject. This chapter considers teaching in general, followed by a discussion of the few incidents I have come across (during years of searching for examples) of possible instruction and learning by older animals.

What is teaching? For humans, it is generally assumed to occur when a more knowledgeable person intentionally conveys information to a less-informed one (usually a child or youth). Higher nonhuman animals (usually mothers) seem to teach their young, too—for example, how to find food and how to beware of predators or, depending on the species, how to *be* a predator. However, for nonhuman animals, it is impossible to prove intentionality. Timothy Caro and Marc Hauser (1992) therefore defined the process of teaching in nonhuman animals as follows:

> An individual actor **A** can be said to teach if it modifies its behavior only in the presence of a naive observer, **B**, at some cost or at least without obtaining an immediate benefit for itself. **A**'s behavior thereby encourages or punishes **B**'s behavior, or provides **B** with experience, or sets an example for **B**. As a result, **B** acquires knowledge or learns

a skill earlier in life or more rapidly or efficiently than it might otherwise do, or that it would not learn at all.

This definition usually means that older animals will be focusing on younger ones, and that the former must not benefit from the instruction. (An elder would benefit, for example, if he or she defeated inferiors in battle, thus teaching them to avoid this oldster in the future.) The researchers noted that an instructor need not be sensitive to a pupil's changing skills, or be able to attribute mental states to others, as these "are not necessary conditions of teaching in nonhuman animals, as assumed by previous work, because guided instruction without these prerequisites could still be favored by natural selection."

Caro and Hauser lumped various case studies under two categories: (1) *opportunity teaching,* such as predator parents toying with prey so their young can practice killing, and (2) *coaching,* where a youngster is either encouraged or punished by an adult for his or her behavior. They note four common mechanisms of instruction that are not necessarily mutually exclusive, and that can be applied to examples of what seem to be teaching by older animals:

1. social learning, such as by imitation or by social facilitation,

2. giving an individual more opportunities to learn,

3. encouraging, and

4. punishing.

IMITATION

An early example of what might be called "teaching by imitation" occurred in Japanese monkeys, even though it was young animals who first began the sweet-potato-washing and wheat-washing behaviors that then spread throughout much of the community. One day a young female dipped her sandy sweet potato in water, thereby washing the sand off before eating it (Kawai 1965). "Her mother and closest peers soon followed, and the habit spread to others. Within a decade, the whole of the population under middle age was washing potatoes" (de Waal 1999).

The same type of scenario—involving what seems like imitation—has been analyzed by Marc Hauser (1988) for vervet monkeys (*Cercopithecus aethiops*), which he studied for many years in Amboseli National Park in Kenya. (He wonders if we are dealing with social learning, social inter-action, exhibition, or contagious behavior.) In this case, the year was 1983, and the "teacher" was not a juvenile, but 14-year-old Borgia dur-ing her last years (she died in 1987). She was by far the oldest in the group of three adult males, four adult females, and three juveniles who made up her troop. Hauser saw her do something astonishing—during a severe drought, she dipped dry *tortilis* pods into a small, otherwise unreachable pool of viscous exudate that had collected in the hollow of an *Acacia tor-tilis* tree. She left the pods in the fluid for about two minutes, then re-moved and ate them. Normally, a vervet monkey does not consume many of these dry pods, but they become important when little other food is available. By leaving them in the liquid for several minutes, they sopped up the exudate (which vervets like), making the dry pods more palatable.

Eight days after Hauser's exciting observation, he noticed Borgia's two juvenile males doing the same thing as their mother. The next day, her adult daughter and another adult female had also caught on. On day 15, the other adult female was on board, as was the dominant male after 22 days. Eating pods coated with exudate became an important part of these monkeys' diet during the drought. However, the other two adult males and the third juvenile (female) in the troop were never seen doing this.

It is difficult to decide exactly what happened here. Was Borgia's act an innovation, or had she seen and remembered it from before the field stud-ies of vervets began in 1977? Certainly the pod-enhanced activity of the oldest member of her group, whether invented by herself or remembered from a much earlier time, was important in helping most members of the troop, including all three of her offspring, survive during the drought. This behavior had become part of the culture or tradition of the small group. Despite some researchers' antagonism to the idea, this learning seemed to be an act of imitation, as it is difficult to believe that the mon-keys would invent such behavior on their own in so short a period of time. Human babies as young as two weeks old are able to imitate the fa-cial expressions of adults looking at them, so it is perhaps not surprising that imitation is also present in young monkeys (Meltzoff 1988).

The case for social learning (or imitation or observational learning) being a method of "teaching" (or at least "training") has also been discussed for domestic sheep and for birds. John Paul Scott (1945) noted that most authors assumed the behavior of sheep was largely instinctive, but experimental work had shown that it could be readily conditioned. Scott believed that learning was an important factor in allelomimetic behavior (defined as any type of activity which includes mutual imitation, something extremely common in sheep) and in leadership. He noted that "it is probable that heredity plays an important part in molding behavior but that it is also modified by learning."

Many scientists have thought that birds, as well as sheep, also behave mostly by instinct, and are poor learners (Diamond 1987; Werner and Sherry 1987). Cocos finches (*Pinaroloxias inornata*), who live on Cocos Island in Costa Rica, use a large variety of feeding techniques to ingest a diversity of fruits, nectars, and insects. However, a close look at individual behavior showed that each bird actually used only one or two feeding techniques, even at varying times of day, months of the year, and parts of its home range. Other individuals feeding in the same bush specialized in different techniques. Juvenile finches apparently adopted a technique from watching an adult foraging from 1 or 2 meters away and then imitating this same procedure, such as gleaning insects from branches, inspecting dead leaves for crickets, or probing flowers to obtain nectar.

Ginger, daughter of the elderly cocker spaniel Rusty, is another example of what seems to be learning by imitation. Both dogs were companions of Carl Marty, who for many years cared for a multitude of wild creatures at Three Lakes, Wisconsin. Sterling North, in his book *Raccoons Are the Brightest People* (1966), reported that during his long life, Rusty lived among many kinds of needy animals, becoming especially good friends with a wild fox (Louie) and a wild raccoon (Snoopy). Eight months before Rusty died, he began to show his puppy, Ginger, "how to play with, and care for, baby otters, raccoons, badgers, foxes, little bears, skunks and even porcupines." Soon, Ginger was "almost as proficient as her father in sensing the mood of each new foundling. Usually within a matter of minutes, and always within an hour or two, Ginger was able to gain the confidence of these woodland playmates"—a remarkable ability apparently "taught" her by her father.

Learning by watching is, of course, also important in people. The great naturalist Sigurd Olson, in his book *Songs of the North* (1987), states that as a matter of pride and masculinity, a junior wilderness guide would never consent to be taught by the senior guide, but he would learn a great deal by watching everything the more experienced man did during canoeing and camping trips.

Examples of the second type of teaching, giving a youngster more opportunity to learn, have been described for two older mothers, an elephant and a fox. The elephant matriarch was annoyed by her infant breaking down the walls of a well she was digging in a dry riverbed (K. Payne 1998). She first gave him a few gentle taps with her foot. When he would not stop, she led him to an empty site some meters away, where she dug a small wallow just for him. She left him playing there in the damp sand while she finished the well she had been fashioning.

By using infrared binoculars, David Macdonald (2000) near Oxford, England, was able to watch red fox (*Vulpes vulpes*) activity after dark. In pastures on still, warm nights, earthworms come partly out of their burrows to the surface, where they can be extracted by foxes. Toothypeg, a nearly 9-year-old female with only one worn canine tooth in her ancient muzzle, was an expert at this. She would grab the end of a worm, hold it taut, and then gradually pull it from its hole, until it snapped out and wrapped around her muzzle, where she could munch it down. Meanwhile, her cub was attacking worms, too, but doing so ineffectually, trying to leap on them as they slipped back into the ground. The cub was finally drawn to Toothypeg's side. When she pulled on her next worm, she held it taut so her cub could take it from her, but he jerked the worm suddenly and snapped it in two. This happened a second time. The third time, only a bit of the worm was visible. Toothypeg held it taut, while patting it with her paw in a light massage, as the tug-of-war continued. Gradually, the worm began to weaken, so that it could slowly be drawn out of its hole. When it was two-thirds above ground, she held it for her cub to snatch greedily. After this, the cub tried again and again to copy his mother's technique, but he was not immediately successful. A week later, he was catching one worm to his mother's four, and after a month he was every bit as proficient at worming as his mother.

ENCOURAGEMENT

An example of encouragement to facilitate teaching has been documented in a group of 12 gorillas (*Gorilla gorilla gorilla*) at the San Diego Wild Animal Park (Nakamichi et al. 2004). Twenty-one-year-old Alberta was interested in the birth of her second grandchild after her daughter, Ione, had neglected her firstborn, often leaving it on the floor so that it eventually was removed and hand-reared by people. When Ione put her new day-old baby, a male, on the ground near Alberta—an unnatural maternal behavior in a species where babies up to 6 months old are almost constantly in body contact with their mothers—the grandmother lifted the baby and handed it back to Ione. Ione tried several times to give the baby to her mother, but Alberta refused to take him. During the next few days, Alberta sometimes held the baby briefly, but then handed him back to his mother. By the fourth day after his birth, Alberta had helped Ione acquire appropriate maternal behavior toward her son. The researchers concluded that the "quality of infant care by the young adult daughter clearly improved during the first four days after birth, and this improvement was at least partly based on her mother's encouragement."

PUNISHMENT

Niall, the 13-year-old female goat, seemed to use indirect punishment as a means of teaching two of her daughters what author Elma Williams (1963) called "herb lore." Niall apparently knew that rhododendron leaves were poisonous to goats, because although she watched the goat Gazelle, whom she spurned, nibble these leaves, she fiercely bunted away two of her daughters when they tried to do the same. Gazelle, who was violently ill the next day but managed to recover, was not a member of Niall's herd, having been adopted separately by Williams and raised by hand. Niall was equally observant when a tree covered with ivy crashed to the ground. She was aware that the berries were poisonous, although the much-coveted leaves were not. She and two daughters brutally charged and fended off all the young goats who crowded forward to feed at the ivy, then gorged themselves on the leaves (but not the berries) when the other animals had dispersed. Gazelle remained a stranger to Niall's herd until her daughter

Sylvan, sired by the father of Niall's grandkids, was born; then Niall finally welcomed both goats into her herd.

There are two examples of older females directly punishing younger ones for their perceived dereliction of maternal duties. Both involve animals in captivity, where they are under constant observation, so such contretemps are likely to be observed and recorded. One case concerned a group of ringtail lemurs (*Lemur catta*) housed at the Duke University Primate Center. One day, a 3-month-old infant climbed up an electric fence, received a shock to her head, and fell to the ground in convulsions (de Waal 1996). Her mother did not see the accident, but her grandmother did. She quickly gathered up the stunned youngster, something she had never done before, and carried her about for 10 minutes before laying her down in a quiet spot. Shortly thereafter, the grandmother again carried her granddaughter here and there. When the mother wandered up, unaware of the past crisis, the infant briefly climbed onto her back. Instead of allowing her youngster to ride for a time, the mother reared up and shook herself violently to unseat the baby. At this, the grandmother instantly attacked her daughter, who then, chastened, allowed the youngster to remount.

This scenario amazed researchers studying lemur behavior. It showed that this primitive primate can recognize when another is in trouble and respond accordingly, as the grandmother had done. By attacking her daughter, the grandmother even seemed to indicate that she knew what the mother *ought* to have done, "precisely the kind of social pressure viewed in moral terms if seen in humans."

Another example of "punishment as teaching" involved a group of female Asian elephants (*Elephas maximus*) kept in the Washington Park Zoo in Portland, Oregon (K. Payne 1998). The prime female, Hanako, had produced three calves, but due to inbreeding, all had serious defects that soon killed them. When her fourth calf, Look Chai, was born completely normal, Hanako had no idea how to care for him; she showed no interest in nursing him, pacing nervously back and forth instead (Schmidt 1992). One elderly cow housed with Hanako, apparently offended by the latter's lack of maternal behavior, attacked her with her head, slamming her into a cement wall. Look Chai was able to nurse briefly from the teats between Hanako's front legs as his mother stood still, suffering from shock. But then she again moved away from him. This time two older females, both experienced mothers, butted her against the wall, and continued to do so for the next few days whenever Look Chai wanted to nurse and Hanako would not stand still. The two aged "aunties" were surely reminding Hanako of her maternal responsibility. Tuy Hoa, Look Chai's grandmother, was so sensitive to his needs that she started lactating herself, and so helped provide him with milk. Unfortunately, Look Chai was taken away from his mother and grandmother when he was only 2 years old, long before he had a chance to learn much from them about being an elephant.

Learning

Barbara Woodhouse (1954) described learning in an old animal. As a compassionate young woman, Woodhouse bought Tommy, a 27-year-old pony with an emaciated body and swayed back; she wanted to "revive his worn-out spirit." By constant repetition and reward, she was able to teach him to climb a ladder, walk on his hind legs, and count. She played cowboys and Indians with him—he raced with her astride to escape from imaginary enemies, and lay down so she could hide behind him to return fire. She noted that Tommy did not know what game they were playing but, rejuvenated, he was willing to go along with anything she wanted because they were so closely attuned.

Sometimes it is impossible even to guess how animals gain experience. In China, herdsmen noticed that older deer (species not given), unlike those who were younger, nibbled at the bitter and astringent bark and roots of fleeceflower (*Polygonum multiflorum*) (Engel 2002). Analysis of its plant parts showed that they had useful properties for elderly people: they increased energy levels and were reputed to reduce hypertension, cholesterol levels, and coronary heart disease. Would young deer surmise the benefits of these plants by watching their elders sample them? When they were older, would they remember to take advantage of what they had seen? And how did deer originally find out about these plants?

RESEARCH INTO LEARNING

It used to be thought that mammals were born with all the neurons in their brains that they would ever have; these cells would be gradually lost as an individual aged. Recently, however, it has been discovered that neurogenesis occurs both in older mice living in enriched environments and in people; even dying human beings have developing neurons present in the hippocampal areas of their brains (Kempermann et al. 2002; Doidge 2007).

It is difficult to visualize how learning might take place in elderly animals in the wild, but research into captive species indicates that it does occur, although old individuals have more difficulty learning than younger ones do, which is also true for human beings. It has also long been known that memory gradually declines with age in both humans and nonhuman primates (Presty et al. 1987). Carrie Dahlberg, in an article for the *Sacramento Bee* (2007), wrote that researchers at the California National Primate Research Center at Davis, where 5000 monkeys are housed, have found that "senior moments" of forgetfulness—when a thought slips the mind or one cannot remember where one put something—occur not only in human beings, but also in rhesus monkeys (*Macaca mulatta*). They are asking questions such as: What effect do hormones such as estrogen have on memory loss? Could a protein inserted in a monkey's brain delay cell death, thus offering promise for research into Alzheimer's disease?

For Java monkeys (*Macaca fascicularis*) studied in captivity, older individuals' ability to process information also declined; this deficit was especially evident in socially subordinate animals who had elevated levels of cortisol, presumably caused by stress (Veenema et al. 1997). In human beings, prolonged or repeated periods of stress also cause elevated levels of glucocorticoid hormones (of which cortisol is one) and impaired cognitive performance (Veenema et al. 2001).

In a study of least shrews (*Cryptotis parva),* older shrews (20 months old) made many more errors in figuring out a complex maze than younger individuals did (Punzo and Chavez 2003). Over a 10-day learning period, their number of errors decreased dramatically, but the oldsters still made more mistakes than their juniors.

SYNOPSIS

The possibility of teaching in nonhuman animals has received little attention thus far in the scientific world. Thornton and McAuliffe (2006) note that "the lack of evidence for teaching in species other than humans may reflect problems in producing unequivocal support for the occurrence of teaching, rather than the absence of teaching." Yet teaching and learning, if intellectually possible, would seem to be important from an evolutionary standpoint. Teaching by older animals, even by imitation, could give young animals new skills that would increase their chances of surviving and reproducing; at the least, teaching could speed up the acquisition of a useful skill, universal to the species, that would otherwise be acquired later on. And of course "teaching"—in the form of doing something important (such as pointing out an unusual food source) that younger individuals will remember for the future—is imperative in passing along a group's experience from elders to the next generation.

IT IS USEFUL TO REMEMBER that schools, universities, instructors, libraries, and the like vastly influence contemporary definitions of teaching and learning. For our human predecessors many thousands of years ago, the reality of teaching was much less formal, and occurred to a large extent within family and community contexts, just as it does (or may do) today in social animals.[1] We should be careful to avoid speciesism, with human beings defining what passes as teaching, which in turn leads to continuing neglect of how teaching and learning occur in nonhuman animals.

CHAPTER 6

Reproduction

REPRODUCTION DIFFERS FOR OLDER ANIMALS and younger ones. (A sample of the behavior of "good mothers" themselves is the subject of chapter 11.) In a colony of adult female langurs, for example, the survival of infants was much better for the six oldest females than for the five youngest ones (Dolhinow et al. 1979). By the time a female is elderly, however, she is often producing fewer and smaller young at longer intervals; that she does so is known as senescence in reproduction. In aging chimpanzees, as in older women, maternal age is directly correlated with negative reproductive outcomes, such as spontaneous abortions and stillbirths in the former, and as well as by birth defects such as Down's syndrome in the latter (Roof et al. 2005). (Not enough older female chimpanzees have been studied to know if such ailments are as significant for them as they are for women.) However, in a few species, when a female is at the very end of her reproductive career, she may give birth to larger young, to whom she gives extra care in what is called terminal investment in reproduction.

There is little exact information on the reproductive importance of older males, given that in the wild it is usually impossible to know which male has sired which young. However, it seems to be generally true that older (but not old) males are more successful in reproduction than younger ones, given the former's increased experience and perhaps larger size (for example, see Sorin [2004] for white-tailed deer; Apio et al. [2007] for bushbuck; Van Noordwijk et al. [2001] for long-tailed macaques; Poesel et al. [2006] for blue tits; Hyman et al. [2004] for song sparrows; Bouwman et al. [2007] for reed buntings; and Best et al. [2003] for southern right whales.)

This chapter first considers examples of increasing success in reproduction as a female gains experience; second, senescence in reproduction; and finally, examples of reproductive terminal investment.

Breeding Success of Older Birds

Research showcases the important role of experience in helping animals reproduce more effectively as they grow older. Young females are notoriously inept as mothers—for example, orcas (see chapter 11), some baboons (such as Vee, who carried her short-lived firstborn, Vicky, upside down, her head bumping along the ground), and many other species—but they become more skilled with experience. The best examples of this effect are in the bird world, where it is possible to collect data from wild species nesting in areas where researchers can document their achievements. They can count the number of eggs in a nest, the number of nestlings, and the number of fledglings (which they usually band for future studies) for each female. The data for females are more accurate than those for males, because it is impossible to know how many offspring a male has actually sired, even in so-called "monogamous" species. The variability among males is also much greater; one male may be far more successful than others at breeding, whereas the females are restricted to raising only one or a few clutches each year.

Often researchers do not know why older birds are more effective breeders than younger ones. In communities of red-cockaded woodpeckers (*Picoides borealis*) in North Carolina, a species at risk because of extensive logging, the production of young increased dramatically with the age of the parents (Walters 1990). One-year-old breeders usually made a mess of their attempts to raise young, 2-year-olds were more successful, and with each succeeding year, the parents' reproduction rate improved, both for males and females.

Another documented example of the benefits of experience—again with researchers not knowing the reason for this—involves communities of Splendid Fairy-wrens (*Malurus splendens*) living near Perth, Australia. These birds are tiny, weighing only 10 g (0.4 oz), the females drab gray-brown but the males a brilliant blue and black; both sexes have the perky tail common to wrens (unrelated) in North America. Older birds are more experienced and successful than younger ones in rearing offspring under poor conditions, and they, along with helpers (often a female's young from the year before), can readily exploit a year of favorable conditions to produce two or even three broods of fledglings (Rowley and Russell 1990). Researchers, over many years, have banded 620 individual birds to discover this. The oldest female in the study area, who was at least 9 years old, bred for eight seasons and produced 10 percent of all the known-age young who themselves became breeders. Both sexes are promiscuous, so we have no idea how successful the various males were as sires, but they lived longer on average than females. The oldest male was 13 years old, while two others were over 10.

Similarly, although we again do not know why, a few Florida scrub jays (*Aphelocoma coerulescens coerulescens*), all fairly old but not in senescence, were far more successful at reproducing than the average adult (Woolfenden and Fitzpatrick 1990). The research area, about 35 territories, was dominated by the descendants of four extremely effective pairs, some of whom bred successfully for nine seasons.

The breeding rate of (related) pinyon jays (*Gymnorhinus cyanocephalus*) is linked to their nesting sites, positioned in pinyon pine trees. Their success, too, improves with experience, so that once again, older birds have a better reproductive record than younger ones (Savage 1995; Marzluff and Balda 1990). In his research on these blue birds—who mate for life and are about the size of a robin—John Marzluff spent six springs climbing 282 trees, some 30 m (100 ft) tall, to observe the nests. He recorded which male belonged to each nest, as well as the precise situation of the nest, and banded each of the three or four baby birds. Later, he noted whether or not the brood survived.

Marzluff found that young, inexperienced jays chose open, sunny spots for their nests. If all went well, the pair picked a similar spot the following year. But if the eggs were taken by predators two or three times, the couple, now 4 or 5 years old, began to prefer lower, more sheltered sites. By the time they were 7 or older, they were still hiding nests low in trees, but these were nearer the ends of the branches, especially if they were built in late winter, to take advantage of increased sunlight. Marzluff noted that "the process by which jays select nest sites appears to involve many forms of learning." Jays learn from their own experience, but perhaps also from other jays. During some warm afternoons, groups of adult jays wing through the colony, pausing to peer into the various nests in the area. Marzluff wondered if they take note of the amount of light and the height of nests associated with large broods of noisy nestlings, the benchmark of success.

Research by Charles Brown (1998) and his wife, Mary, involved banding 80,000 individual birds and studying a variety of colonies over 14 summers. They discovered that in Nebraska, older cliff swallows (*Petrochelidon pyrrhonota*) were more likely than younger birds to nest successfully in small (rather than large) colonies. The reason for this seemed to be that in large groupings, fleas and swallowbugs (similar to bedbugs) were more numerous and therefore more deadly, sucking blood, spreading disease, and, if there were enough of them, killing nestlings. The couple also noticed that the oldest birds they netted, 11 years old, looked no older than the young adults.

Senescence in Breeding

The senescence hypothesis of reproduction predicts that as mammals and birds age, individuals gradually become physically weaker, lose condition, and produce and raise fewer young (Weladji et al. 2002). This has been well documented in many species. For the very oldest scrub jay breeders in the study mentioned above, those birds referred to as "in senescence" had significantly fewer eggs in their clutch, lower fledgling success, and lower fledgling survival than younger parents. In at least 15 species of apes and Old World monkeys, aging females are reported to have a declining number of young, with longer intervals between births than younger animals (Paul et al. 1993); a number of these older females experience menopause and, in the case of chimpanzees in the Mahale Mountains, up to nine more years of healthy life for one quarter of the elderly females (Nishida et al. 2003). Some of the oldest female Columbian ground squirrels have senescence in reproduction (Broussard et al. 2003), as do tree swallows (Robertson and Rendell 2001). Bird species may also exhibit senescence in other ways that affect reproduction, such as a decline in offspring quality (for barn swallows, see Saino et al. 2002) and changes (dispersals) in breeding sites (for blue-footed boobies, see Kim et al. 2007). However, in a recent study of reproduction in deer (DelGiudice et al. 2007), there was "no evidence of senescence relative to fertility and fecundity in adult female white-tailed deer up to fifteen and a half years old."

Another recent study documented the reproductive success of mountain gorillas (*Gorilla gorilla beringei*) over 38 years in the Virunga volcanoes region of Rwanda, Uganda, and the Congo (A. Robbins et al. 2006). The data include all the animals watched by Dian Fossey (in her research begun in 1967), as well as by many other researchers since, for a total of 214 mothers and their young. Miscarriages occurred more often in older rather than younger females. The birth rate was lowest for females over 40, the oldest group, as was the number of infants who survived to 3 years of age. There were longer delays in resuming reproduction after an infant died. And there was no evidence of an extended post-reproductive lifespan. Robbins and his coauthors report that "these combinations of results seem most likely to reflect changes in the physical condition of the mother, rather than her level of investment and experience." That mountain gorilla females had a post-reproductive lifespan of only 1 to 3 percent of their total lifespan reflects the reality of their daughters leaving their natal troop when they reach puberty, so the grandmothers have no biological descendants on whom to lavish attention. For primate species in which daughters stay in their natal troop, Japanese monkeys have a post-reproductive lifespan of 9 percent of their total lifespan, and baboons, 16 percent. (Women have a post-reproductive lifespan of 20 to 40 percent of their total lifespan.) Older Japanese monkeys had estrus cycling and sexual activities similar to those of the young adults, except that their estrus periods were shorter, so the decline in their fertility does not seem to be due to menopause (Wolfe and Noyes 1981). It may instead be caused by deteriorating health, tumors in their reproductive organs, or increasing chromosomal aberrations. Older bonnet monkeys (*Macaca radiata*) also had sexual behavior qualitatively similar to that of younger bonnets, although during a period of stress when new social groups were being reorganized, they suffered a far more drastic drop in reproduction (Jensen et al. 1980).

The data for wild mountain gorillas, however, differ from the physiological data for western lowland gorilla females (*Gorilla gorilla gorilla*) held in zoos across North America. In general, the lowland females live a long time (Atsalis and Margulis 2006). Their maximum longevity to date is 52 years, with reduced reproduction beginning from age 37. For these older females, their reproduction is characterized by both perimenopause and menopause, so that they may experience a post-reproductive lifespan of more than 25 percent of their total lives. One captive female, who was older than 40, still had progestogen peaks (based on hormones present in her feces), which showed a regular and close coincidence with monthly sexual behaviors present in younger females (Atsalis et al. 2004). Fertile women have similar peaks in their menstrual cycles following ovulation.

Terminal Investment in Reproduction

The evolutionary hypothesis of terminal investment in reproduction predicts that parents/mothers should invest more in their young as their own potential for future reproduction diminishes (Clutton-Brock et al. 1984). This hypothesis is antithetical to that of senescence, and indeed the two could have the effect of canceling each other out if both were in effect, which makes research on this topic difficult. In addition, the environment plays a part in reproductive success, with older parents/mothers and their young, like other "families," surviving better in good times than in bad.

Only a few species (mostly large game animals of interest to sports men, and primates of interest to evolutionary anthropologists, psychologists, and biologists) have been studied long enough so that the age of many older mothers, along with their reproductive histories, is known. Thus it is premature to make any sweeping generalizations about the hypothesis of terminal investment in reproduction. We can, however, consider a few sometimes puzzling examples of it in disparate species: red deer, reindeer, moose, Barbary macaques, California gulls, and collared flycatchers.

UNGULATES

One of the earliest studies on the topic was the breeding behavior of red deer living on the Island of Rhum in Scotland. Zoologists from the University of Cambridge researched the changing deer populations for many years, recognizing individuals by their body shapes, coloring, and facial features; they also used ear tags and colored expandable collars on some deer for identification (Clutton-Brock et al. 1982). They followed the life vicissitudes of a population (in 1979) of 149 hinds and 135 stags. Survival over the winter was high among calves born to hinds age 3 to 6, lower among calves of 7- to 9-year-old prime-age hinds, and high once again among calves of older females (age 10 to 13).

This discovery amazed the researchers, because the older deer were in poorer condition than their younger sisters. However, when the men timed the winter nursing bouts of calves belonging to hinds of different ages, they found that these bouts were longer in the oldest age group, indicating that these hinds, despite their lower body weight, were sharing more of their resources with their young. From an evolutionary perspective, this finding implied that prime females—as opposed to older females—could afford to let a calf die in a hard winter, because they had a good chance of producing more progeny in the coming years. This has been called the evolutionary restraint hypothesis, which predicts that young mothers should have a relatively low investment in their current offspring, and a relatively high investment in their own maintenance and continued growth (A. Robbins 2006).

Could it be true that the calves of aging females were more hardy than those of prime females? The researchers also considered other possible explanations. Could the production of blue-ribbon calves by these older animals be related to the fact that a proportion of elder females failed to breed every year? No, since blue-ribbon calves were produced by hinds that both had and had not bred the previous year. Perhaps the aging females failed to breed the next year? No, because both the pregnant and the barren elderly hinds later produced similarly healthy offspring. Perhaps, from a comparative perspective, lightweight calves born to older hinds died in the summer, whereas similar calves born to prime females lived through the summer but died in the winter? Again, the data showed that this was not so. Red deer females indeed seem to exhibit terminal investment in reproduction.

More recently, statistical studies on this evolutionary hypothesis have been carried out on 1656 semi-domestic female reindeer (*Rangifer tarandus*) living in southcentral Norway (Weladji et al. 2002). All the reindeer from this herd had been tagged as calves, so the exact age of the females was known. The weight of each calf was obtained at 2 months of age, as well as the weight of each calf's mother.

Their results showed that the weight of the calves peaked for 7-year-old females. The young progressively decreased in weight for females 8 years and older, which supported the hypothesis of senescence. The researchers concluded that either the terminal investment hypothesis was wrong, or that reproductive costs increased with age. They noted that other recent studies also corroborated the senescence hypothesis, but not the terminal investment hypothesis, for ungulates such as female bighorn sheep and roe deer. They also questioned the existence of terminal investment in reproduction for the red deer of Rhum, because the actual weights of mothers and calves were not considered in the research.

The reproductive history of male ungulates is far more difficult to study than that of females, because we know which youngster belongs to which mother, but not who his or her father is. Even so, researchers have analyzed male reproductive behavior in several species. During the fall rutting season, prime red deer males fight each other fiercely for the privilege of herding a group of females, with whom they will copulate. These groups are sometimes called "harems," but this is a misnomer, as the groups' composition is constantly shifting and changing from day to day.

In their study of this species, Timothy Clutton-Brock and his colleagues (1982) found that most stags over age 11 failed to defend such female groups, and reverted to wandering around among the groups herded by prime males, seducing an occasional hind where possible. Eleven is also the age at which these males no longer take part in fights, and when both their condition and weight typically begin to decline. Older (and young) males herded groups of females later in the year than did the prime males. As he matured, one successful stag, SAGY, defended "harem" groups progressively nearer to the onset of the peak rutting period; as he aged, his rutting period again occurred later, when competition among the males was less fierce. Stags can lose as much as 20 per cent of their body weight during their rutting period. Success in breeding for individual older males was estimated (this was before the age of DNA testing) to range from no calves fathered at all to as many as 25 surviving to be 1 year old. Aging stags have difficulty mating with females, because they are not strong enough to defeat prime males in battle, so there is no possibility of them having evolved a terminal investment strategy for reproduction (Mysterud et al. 2005).

Moose (*Alces alces*) behave very differently than red deer, because they are basically a solitary rather than a social species. Researchers in Sweden placed radio collars on 127 free-ranging cow moose (after using a helicopter to dart the animals with a tranquilizer), and then checked their reproduction rates over an extended period of time (Ericsson et al. 2001). These females produced 351 calves, 211 of whom were weighed shortly after birth. The scientists therefore knew the age of the mother; the age, weight, and number of her young (one, two, or three); and their possible mortality.

The researchers found a fairly constant litter size for animals aged 5 to 12, followed by a rapid decrease to the end of their reproductive period, at age 15, due to senescence. Older mothers gave birth to heavier offspring, which was needed if their calves were to achieve the same summer survival rate as the offspring of younger females. Thus aging moose, like elderly red deer females, increased their reproductive investment in their young. Even so, older moose had calves who were more likely to die in summer than other calves.

Bull moose never encounter females in groups. During the fall rut, males respond to the moaning cry of females in estrus, mating after searching for them in the forest. Bull moose shot during the annual fall hunting season in Norway are carefully managed for maximum yield, with hunters required to record the sex, date of killing, locality, and body weight of each animal. Mandibles are collected, so that each kill can be aged by tooth eruption patterns and tooth sectioning. Therefore, the weights of males shot both before and after the rut can be compared.

In a study in Norway, researchers assumed that since bull moose do not eat during the rut, their weight loss during this period is a measure of their reproductive effort (Mysterud et al. 2005). The carcass weights of 9949 bull moose, aged 1 to 21, showed that the reproductive effort increased with age up to age 12, well past the prime age of about 6, "the first evidence consistent with the terminal investment hypothesis in male mammals." The researchers noted that when a moose population was female-biased (owing to previous over-harvesting of bulls), the older males became senescent earlier than they did if there was an even sex ratio, presumably because of their increased exertion during the rut.

PRIMATES

Research on terminal investment in reproduction for older female primates is difficult to carry out in the wild, but is possible under semi-captive conditions. One 11-year research project studied 207 Barbary macaques (*Macaca sylvanus*) living in a large outdoor enclosure in southwest Germany. In this group, older mothers weaned their infants significantly later (by nearly two months) than did young mothers, which explains in part their longer interval between pregnancies. These elders also spent considerably more time hanging out and playing with their young (Paul et al. 1993). The offspring of aging females showed the highest survivorship in the study. Therefore, older Barbary macaque mothers did contribute more than usual to their last offspring, as the terminal investment hypothesis predicts. These seniors were not the most reproductively successful, however, because the high survival rate of their infants did not compensate for their declining fecundity. Among the oldest macaques (in their mid-twenties), estrus cycles stopped three or four years after the birth of their last young, and several years before their deaths. However, a study of female reproduction in a related species (Japanese monkeys) failed to support the maternal investment hypothesis (Fedigan and Pavelka 2001).

BIRDS

Large mammals who reach adulthood tend to live a long life before becoming senescent and dying. For example, Barbary macaque females who survived to adulthood had nearly a 60 percent chance of living to age 20 (Paul et al. 1993); one quarter of the adult chimpanzees in the Mahale Mountains in Tanzania died of old age (Nishida et al. 2003); and pronghorn antelope who reached adulthood had nothing to fear from predators because of their speed (Byers 2003).

In contrast, the survival graph for adult birds (all fairly small) generally has a constant downward trend, as these birds are continually picked off at random by predators such as hawks, eagles, weasels, and foxes. With only a few individual birds reaching senescence, "discriminating the ages of breeding birds is much less important than it would be in life-history studies of many other organisms, such as mammals or fish" (Winkler 2004). The terminal investment hypothesis is probably not as common in birds as it is in mammals. Indeed, in the enormous 1200-page *Handbook of Bird Biology* (Podulka et al. 2004), the phrase "terminal investment" is not in the index.

It may be uncommon, but terminal investment in reproduction has been documented in a few bird species, including California gulls and collared flycatchers. Bruce Pugesek (1981) studied California gulls (*Larus californicus*) nesting on an island near Laramie, Wyoming. Since 1959, more than 1000 gulls had been banded there (in a population of over 4000), so that many individuals could be recognized. Pugesek carried out his observations on a sample of 59 nesting pairs: about one-third were young gulls (3 to 5 years old), one-third middle-aged (7 to 9), and one-third elderly (12 to 18). Numbered stakes were driven into the ground near each nest, so that an observer could recognize each pair from a nearby observation tower. If the birds were away from the nest, their numbered bands helped identify them, as did marks added on some of their wings. Among the data collected were: the number of surviving chicks fledged by each pair of gulls throughout the breeding season; the amount of time that the gulls spent foraging for food to feed their young (which was assumed, from observations, to be about the same amount of time each parent spent away from the nest); and defense activity by the parents regarding their young and their nest area.

Pugesek found that the average number of offspring fledged in a year was much higher for the older gulls (1.5 young) than for the middle-aged pairs (0.8) or the young parents (0.76). During the nesting period, the parents took turns foraging for food and sitting on the nest. This latter activity ensured the safety of the eggs or nestlings from predators and other gulls. However, sometimes the sitting/guarding bird did not wait for its partner to return before flying off on its own. Young gulls were more likely than older birds to leave their nests unprotected in this way, and to do so for the longest periods. The older birds were not only better guards, but also more assiduous in and more experienced at foraging. They fed their chicks more often than did the other gulls, and fed them for up to five days longer. The older birds also defended their territory more than the others did. In both foraging and defense activities, the elders put themselves at greater risk of injury than if they had remained resting within the colony.

Pugesek attributed the increased reproductive success of older seabirds to experience and social status. According to evolutionary theory, older gulls expend more energy on their nestlings than younger gulls, and have better success in raising them, even though these activities may weaken aging parents and jeopardize their return fall migration to California. This extra effort is less important for young birds, who have a rosy reproductive future ahead of them. Therefore, Pugesek observed, "long-lived animals may be selected for reproductive restraint while they are young and increasing reproductive effort as they grow older. Increased reproductive success with age reported for many species of seabirds may be a result of similar selective pressures."

Another bird study on terminal investment in reproduction was carried out on collared flycatchers (*Ficedula albicollis*) nesting on Gotland, a Swedish island in the Baltic Sea (Pärt et al. 1992). Although these birds migrated south in the fall, they always returned to the same area in the spring, where they were caught in mist nets and weighed. By examining their nests, the researchers recorded clutch size, nestling mortality, and the number of fledglings, which they banded.

In 1988 and 1989, they examined the rates at which birds of different ages fed their young. The study compared fifteen pairs of older (at least 5 years old) and younger but experienced (2- to 3-year-old) females, each matched for the same date of hatching and clutch size, and each with a mate to help with the feeding. The older females fed their nestlings more often and lost more weight during the nestling period than did the younger females, but, surprisingly, the weight of the nestlings was the same for the two groups. Further analysis showed that the mates of the younger females helped more with the feeding than did those of the older females. The authors theorized that there was a conflict of interest between the older females and their usually younger mates. It makes evolutionary sense for a male to decrease his feeding effort if the female has increased hers, because the nestlings are likely to survive, and the male will save his energies for further reproduction in the future.

THIS CHAPTER LOOKED AT REPRODUCTION in general. The next chapter will consider the reproduction of older animals in the context of their social ranking within their group. Evolutionary theory presumes that dominant animals should be more successful in producing young than subordinate ones, but this is not necessarily the case.

CHAPTER 7

Successful Subordinates

W E USUALLY THINK OF SUBORDINATE animals as losers, but they can also be successful—by living longer than others, by producing more progeny than do dominant individuals, or both. Dominance and subordination, which run through most animal societies, affect older animals in various ways. Dominant individuals are those who can defeat other animals in battle and, by doing so, obtain more and better food, the best resting sites, and superior partners with whom to mate. Hierarchy in societies is important because it enables each animal to know his or her place in the ranking, thus avoiding constant conflict. For most animals who live in mixed-sex groups, males are dominant to females—not because of their sex, but because of their greater size. When females are bigger than males, they are the dominant sex, as in the four Hs: hawks, hares, hamsters, and hyenas. Ringtail lemur females are also dominant to males, and the captive older chimpanzee Mama was dominant to all the other males and females in her group. Where male and female adults live largely separate lives, each sex usually has its own hierarchical ranking. In most animal societies, individuals gain dominance (if they do so at all) after they reach their maximum size and strength; they lose it as they become less strong or grow frail with old age. A few societies function so that an individual born to dominance (for pronghorn antelope) or to dominant parents (for common baboons, vervet monkeys, and Japanese monkeys) retains this dominance throughout his or her life, even into old age.

Dominance is especially important for males, because it affects their ability to reproduce. Whereas virtually all female adults continue to breed regularly until they grow old, males vie with each other for mating opportunities; some may have great success in breeding, while others have no success at all. In the extensively studied red deer of Rhum, there was a dominance ranking among the females, but rank did not correlate with reproductive success (Clutton-Brock et al. 1982). In contrast, the higher-ranking stags had greater reproductive success than did their male inferiors. In one-male groups of animals, as the dominant alpha male grows older, he is usually defeated by a younger male, and so (unhappily for him) becomes subordinate. This happened to Yeroen the chimpanzee, Solomon the baboon, and other individuals described in chapters 8 and 9. However, subordinate older males may still find ways to mate with females, as elderly baboons do by becoming Friends with young adult females (see chapter 13).

Dominance is important for females, because it allows them to have the best of everything: food, water, and resting sites. It also bestows superior rank on their offspring. For example, if one of their youngsters is in trouble or even seems to be so, the high-ranking mother will leap to his or her defense.

Hormonal studies have not clarified why individuals belong to one rank rather than another. In many social species of birds and mammals, reproductive rates are lower for subordinates than for dominants, and in some species reproduction in subordinates is completely suppressed, such as in wolf packs (Creel 2001). Early research showed that glucocorticoid (GC) secretion increased in captive animals who lost fights and, because GCs can suppress reproduction, it was thought that chronic stress might correlate with reproductive suppression. More recently it has been found that for some species of social carnivores in the wild, dominant individuals have elevated GCs more often than do subordinates (Creel 2005).

This chapter deals with several aspects of dominance and subordination. First is a description of interactions between the older baboons Leah and Naomi, to show how annoying rank order repercussions, for apparently no good reason, can be to individuals in everyday life. Next is a discussion of how, in a strictly lifelong hierarchical society—such as that of vervet monkeys—an older individual (Marcos) has occasionally been able to rise in rank. Finally, there is information on individuals from three species (pronghorn antelope, mountain sheep, and wolves) who have lived a long and sometimes productive life despite always being subordinate. Intuitively, it seems that dominant animals are much more likely than subordinate ones to reach old age, because of the deference demanded by them. By living until they are elderly, it seems logical that such animals have the best chance of producing many progeny to carry on their inheritance, the benchmark of evolutionary success. But the correlation with rank position in the hierarchy and superior reproduction is not always straightforward.

Rank Order Repercussions

An irritating consequence of being subordinate is that it can affect every day of one's life. Consider the old olive baboon (*Papio anubis*) females Leah and Naomi. Baboon males in a troop do not retain a lifelong dominance ranking, but females do. Males may be dominant in their group at some point in their prime, but individuals sporadically join or leave a troop, so there is no permanent male long-term hierarchy. In contrast, female baboons born into a troop stay there all their lives. They have a rigid hierarchy, one to another, that does not change to any extent from the day of their birth. Leah, at 25 the oldest female in Sapolsky's (2001) troop, inherited her high rank from her mother. Since she ranked above Naomi, who was nearly the same age, she harassed her continually. If Naomi settled down under a shady tree, Leah soon ambled over to force her to move, which Naomi did. Another piece of shade for Naomi, the same scenario. (Baboons here lived like royalty, because they could find enough roots and tubers to feed themselves in four hours a day, leaving eight hours to relax and annoy other baboons if they were so inclined.) Hierarchy in common baboons means that one female gets the best out of life, while another gets the dregs, no matter what either does. Male life is more democratic, because dominance depends not only on heredity, but also on one's personality and aggressiveness (Sapolsky 1994).

Elevation in Rank Order

Vervet monkey females, like female common baboons, have a strict hierarchical system. In vervets, lower-ranked individuals are recognizable because they spend more time grooming higher-ranked individuals than the other way around. The males, because they immigrate into a troop, have a social organization that is much less stable over time. Dorothy Cheney and Robert Seyfarth, in their book *How Monkeys See the World* (1990), reported that from their four-year study of the behavior of vervet monkeys (an animal about the size of a cat) in Kenya's Amboseli National Park, females changed rank at the rate of one rank change per female every 10 years. By comparison, males changed rank seven times as often—rank depending not on tradition, but on an individual's size, strength, age, and aggressive tendencies.

Occasionally, a female vervet monkey with few close kin is toppled from her high rank by the persistent aggressive challenges of a larger, lower-ranked female. Sometimes such a change can happen more by chance, as it did to Marcos, who lived to be 16 years old. In group B, Marcos ranked sixth out of seven females. However, between 1977 and 1981, she gave birth to three surviving daughters who inherited her ranking. The female who ranked just above her, Duvalier, produced only two such daughters. In 1981, Duvalier was eaten by a python, but her two daughters retained their status above Marcos and her daughters for the next two and a half years, until one of Duvalier's two daughters died. Meanwhile, Marcos had produced two more daughters, and her oldest daughter also had two daughters. Marcos's matriline was winning the battle of the cradle, as it were, and with daughters rather than sons. Marcos and her progeny now had little trouble rising in rank above the lone Duvalier female.

Years passed. By the time she was 13, Marcos had more offspring (nine), and especially more daughters (seven), than any other vervet female in Group B. Over the years, because of her large family, other females had taken to grooming her more than they would usually groom a female with her low ranking; a large family meant that she would be able to recruit more relatives to her aid in any conflict. Gradually her status rose. By 1986, Marcos was the second-ranking female. She, along with her family, were subordinate only to Amin and her daughter Aphro. The following year, when Aphro's daughter and Amin both died, Marcos became the alpha female almost immediately, groomed more than any other female in the group. Her many progeny shared her high rank. She had achieved (or lucked into) an almost impossible goal for herself and her family—moving from rags to riches in her lifetime, and becoming dominant so that she and her family could now boss all the other vervet monkeys, and choose the best of everything for themselves.

Subordinate Yet Long-Lived

PRONGHORN ANTELOPE

Female pronghorn antelope society is also incredibly hierarchical. These antelope (*Antilocapra americana*) spend their entire lives either butting their compatriots into submission (if they are dominant to them), or being pushed about (because they are subordinate). But it is not necessarily the dominant animals who live to old age and produce many young. John Byers, in his book *Built for Speed* (2003) about his research in the National Bison Range in Montana, described an older subordinate female, GY. She was the most successful of her age class, and was still alive and reproducing after all her bullying sisters were dead.

Byers began his studies by fastening ear tags on some individuals, marking others temporarily with paint balls shot from a pellet gun, and making notes on head and throat coloration and other unique details (such as horn size and horn orientation), so that he could recognize every animal through his binoculars. Over a 20-year period, he noted down the many social activities within two kinds of groups: females with young, and bachelor males. (These groups are not really herds in the usual social sense of the word, but rather individuals who happen to be in the same place at the same time.)

Pronghorn antelope females, who weigh about 45 kg (100 lb) each, are amazingly robust and prolific, producing twins yearly, each weighing about 3.6 kg (8 lb). (If the average fertile woman were as productive, her yearly newborn would weigh 8.6 kg [19 lb]!) Weight is important, because a newborn pronghorn immediately starts nursing from his or her mother, gaining 0.23 kg (0.5 lb) of weight a day. Fawns who are born early in the spring, even a day before other fawns, are bigger and therefore dominant to their fellows. If they spot a smaller fawn nibbling at particularly succulent grass, they march over, force that fawn to move off, and begin feeding there themselves. They are copying the behavior of their mothers, who spend much of their time harassing females inferior to themselves. GY and her twin were born late in the season, which explains their inferior status. Once a fawn's dominance ranking is determined, it continues for life, both for females among females, and for males among males in the bachelor groups.

Byers was curious about why dominance and subordinate interactions were so common, yet could eventually benefit inferior animals—such as GY—more than those dominant to her. He studied this question for two summers. Together with a helper, Byers recorded every dominance interaction that occurred anywhere in a group—the name of the initiator, the name of the recipient, what happened, and who won. In every case, a female who was dominant to another female won every interaction between them. If a female joined a group of females, she immediately looked around to see whom she could dominate, and then proceeded to approach that one with a bold stare or lowered head, as if to butt her. The other female, such as GY, immediately moved away.

Byers was amused when he watched pronghorn females preparing to lie down after they had grazed for most of the morning. Each looked about, wondering who would lie down first. When one did, a more dominant female would usually threaten her, so that she had to stand up again while her superior took her place. The displaced female then challenged another subordinate, who would move away, and a chain reaction ensued. After much milling around and forcing up, at last they would all be recumbent, with the subordinate females usually at the edge of the group.

It was this feature—of subordinate animals usually being peripheral to the group—that interested Byers. Today, the main predators of pronghorns are coyotes and eagles, who by chance come across young fawns hidden in the grass, waiting for their mothers' return. Juvenile and adult pronghorns are too alert and speedy—sprinting at nearly 96 km/h (60 mph), and cruising for several miles at 70 km/h (43 mph)—to be caught by any predator. But some 10,000 years ago, before the ancestors of cheetahs, lions, and hyenas were exterminated from American grasslands, being at the edge of a group made individual pronghorns vulnerable to these predators. This is true today for other species of antelope that are preyed upon by cheetahs in the savannas of Africa. The very fact that pronghorns gather in groups, rather than being solitary, is probably because of this past history of predation. The more eyes there are to detect danger, the more likely the gathered members are to notice the presence of predators and flee.

Old GY survived a lifetime of being bullied. She had to rest at the periphery of her group, but luckily nowadays that would not make her more vulnerable to predators. She could never be sure she would be allowed to eat a newly-found, tasty patch of vegetation before she was shunted aside by one of her group-mates. She could never be certain of lying down to rest before the other animals were settled. She may have been inferior in rank because of an accident of birth (the date she was born), but she was one tough cookie. Unlike other pronghorns in her age class, GY was able to survive through summer droughts and winter storms, produce two fawns year after year, and roam the glorious wilds of Montana.

MOUNTAIN SHEEP

Among mountain sheep (*Ovis canadensis*), dominant males tend to die young, as Valerius Geist discovered in his extensive research in western Canada. In his book *Mountain Sheep: A Study in Behavior and Evolution* (1971), he reported that rams with large horns reach breeding status earlier (at age 6 or 7) than those with smaller horns, and that the former have very good breeding success in their prime. They spent the fall rutting season battling with other males for the privilege of mating with the estrus ewes. (This is not hard to believe for those who have seen a fight on film between two males, the rams smashing heads against each other with echoing crashes, and snatches of the "Anvil Chorus" hovering as suitable music in the background.)

These feisty males were so exhausted and emaciated at the end of the rut, both from chasing rivals and females and from eating little because of these exertions, that some did not survive the following winter. Of the 40 rams found dead in the field, those who died below the average age of death (about 10 years) had better horn growth than those who died after the average age. The older, subordinate mountain sheep males may not have fathered many young (or maybe by chance they did, bit by bit, since they lived longer than their aggressive cohorts), but at least they had a longer, less stressful life.

Geist was not able to learn much about older mountain sheep females who lived year-round in their own groups, because the segments of their small horns could not be accurately counted, unlike those of the large-horned males. A female's age could only be known if she had been caught and tagged when young. Older mountain sheep ewes were followed by their yearling young, and their newborn lambs if they had them, but they did not lead the larger groups. This organization is reminiscent of that of the domestic sheep studied by John Paul Scott (1945).

Among wolves, as among pronghorn antelope (such as the female GY), it is not necessarily the alpha animals who live the longest. Subordinate individuals can also outlast their peers, as Lakota did. He was a member of a wolf pack in a huge enclosure in Idaho, observed for many years by photographer Jim Dutcher and his co-author wife, Jamie. In their book *Wolves at Our Door* (2002), they record that Lakota was one of four pups born in 1991 to the founders of the Sawtooth Pack. Unlike his siblings, he was unaggressive and shy around strangers, personality traits that contributed to him being chosen as the omega individual, the one to be attacked and scorned by all the other wolves. During social interactions, he spent most of his time lying on his back, whimpering in surrender to more assertive individuals, especially the alpha male Kamots. Although Lakota was a huge wolf, larger than most of the others, his skin was riddled with scars and bumps where the other wolves had bitten or scratched him. Lakota had a wonderful howl, musical and sad, but when he joined his voice to that of his pack mates, they sometimes took offense and drove him away, his tail between his legs.

Eventually the Dutchers were required to give up their connection with the wolves, and move them to a new area. This caused the pack to become dysfunctional, especially after their two familiar human caretakers resigned from their positions. Kamots died, the beta male was removed to a separate pen because the younger wolves were tormenting him, and the alpha female was so abused by her daughters that she managed to escape from the enclosure. However, Lakota lived on, by now an elderly animal, the mildest and gentlest wolf the Dutchers had ever known. Surprisingly, he was finally relieved of his omega status, when it was conferred instead onto a younger animal. Lakota at last became the "wise old man of the pack, for in his life he had seen and endured the most."

IN GENERAL, DOMINANT INDIVIDUALS HAVE more descendants than subordinate ones, but this is not always true. Subordinate individuals, especially females, may also have stellar careers, living a long life and often producing many offspring.

CHAPTER 8

The Fall of Titans

SUCCESS IN THE ANIMAL WORLD involves producing as many young as possible. In social species, usually being the highest-ranked individual in the group—an alpha male (patriarch) or alpha female (matriarch)—makes this feasible. But the aging process means that there are always younger, eventually stronger animals growing up to replace their elders. This chapter describes the descent of some alpha animals as they lost their high status to younger rivals: redtail monkeys; chimpanzees such as Mike and Goliath; hamadryas baboons Padorn, Rosso, and Admiral; olive baboons such as Solomon; Pan and Jupiter, the Japanese monkeys; Diva, the female ringtail lemur; and Leo the lion.

Redtail Monkeys

In most species, when an alpha animal is being displaced, the rest of the group knows who is in control at any one time. Occasionally, male displacement can be a slow process, as it is for redtail monkeys (*Cercopithecus ascanius schmidti*). Over a period of many months, Thomas Struhsaker (1977) studied redtails inhabiting the Kibale Forest of Uganda. They live in groups of about 35, each with a single patriarchal male. If this male is challenged by a male of about equal strength, the winner may be in doubt. In part because of the group's large size, neither male is forced out of it, yet neither has exclusive access to it, either. The older, longtime alpha male tries his best to avoid or to minimize conflict with the challenger, in order to remain with the group and mate with females for as long as possible. In this state of transition, solitary male adults, who usually hover around the edge of the group, infiltrate it. They themselves try to mate with females, and scrap with other interlopers who have the same ambition. Eventually, the aging alpha male is ousted by the challenger, who begins his own more stable reign.

Chimpanzees

Jane Goodall's definitive book, *The Chimpanzees of Gombe* (1986), often describes the activities of specific animals within this species (*Pan troglodytes*). Thus we come to know Mike and Goliath, who were among her oldest acquaintances at the time of their deaths. Both were alpha males in their prime who, when older, fell to the bottom of the dominance order, the position where they had started out many years before.

Mike, by using his wits, became an alpha male as a youth. One day he charged toward his superiors (who had ignored him up until then) several times, banging together empty kerosene cans that made a terrific racket. This aggressive stratagem scared the dominant animals into placating him, and soon won him the position of alpha male. By careful plotting, he held on to this status for over six years. For example, he kept out of the way if it looked like he might have to fight, and he threatened others with a variety of artifacts such as tripods, chairs, and tables (before these were removed), in order to enhance his frequent displays of aggression. But toward the end of his reign, his displays were largely bluff. When he was finally deposed at age 32, he looked elderly, with worn tooth and sparse hair. He did not seem to care much that his status was reduced. Soon he was showing submissive behavior to even the lowest-ranking adult males. During his last four years, he became even more of a loner, spending time only now and then with old Hugo as a companion. Goodall writes that he was a master "of reasoned thinking and skilful tactical and social manipulation."

Unlike Mike, who deposed him from the alpha male position and who himself later plummeted suddenly to the lowest rank, Goliath managed to hang on to his high status for years after his dethroning. Goliath lost alpha status to Mike when he was 26 (after the episode of the clanging kerosene cans), but his level of aggression was so great that his high ranking continued for another five years. Then he became sick and emaciated, which undermined his aggressive spirit. From that time on, he, too, was near the bottom of the dominance ladder. When his group divided into two, he unwisely chose to join the breakaway group, and was killed by his former colleagues four years later, at the advanced age of 37.

Less well-known than Goodall's chimpanzees of Gombe are those of the Mahale Mountains in Tanzania, where many of the zoologists have been Japanese. Their published work is often less anecdotal than that of Goodall. The Mahale researchers found that with increasing age, chimpanzees largely withdrew from social interactions within their group, their individual behavior depending upon such things as the presence or absence of offspring, health, non-kin social relationships, and reproductive state (Huffman 1990). Old males remained the more social sex, often staying within the core group after they lost dominance, and sometimes forming a duo with a younger male, perhaps a grooming partner (Kawanaka 1990). Females tended to be less sociable than males, although one remained with her son, a prime male, and a few associated with an offspring or with an orphan. Like males, they were more often groomed by others than vice versa. Old individuals may have been subordinate, but they had a unique status in chimpanzee society, not only luxuriating in being groomed, but also well-tolerated by younger animals—to the extent that an old male could threaten a younger, stronger male without that male retaliating (Huffman 1990).

Hamadryas Baboons

Hans Kummer (1995) wrote about the behavior of hamadryas baboons. After studying them in captivity in the Zurich Zoo (see chapter 9), he wanted to observe them in the wild, which meant the arid regions of Ethiopia. Unlike savanna baboons, where it is the males who change troops to prevent inbreeding, in hamadryas bands it is the females who transfer. They do so when young, and are forced to remain with a male, in a harem-type structure, for the rest of their lives.[1]

Kummer became the mentor of Jean-Jacques Abegglen (1985), who observed the decline of three old alpha males (Padorn, Admiral, and Rosso) in a baboon band that visited Cone Rock in Ethiopia. All of them were defeated in battle by their followers—beware the Ides of March. The "magnificent" Padorn was vanquished in a conflict with his follower Spot, which left him with bloody wounds, an injured arm, facial scars, and the loss of his two adult females. Within three weeks, Padorn's face turned from brown to ash-gray. Two of his peers, Rosso and Admiral, threatened Spot on his behalf, but to no avail.

A year later, these two older males were themselves wounded in fights with their closest followers, and overthrown. Admiral lost all nine of his females and seven daughters, while Rosso forfeited three of his four females. The coveted adult females were now attached to the new prime young males (none with visible wounds from fighting), except for two females who disappeared entirely from the band. Within one year, these three dominant "harem" leaders had lost all their adult females to followers (who were not the leaders' sons), with the exception of Rosso's one female (Abegglen 1985).

The three old males all had worn canine teeth and were slowing down with age. With the loss of their "harems," however, their aging accelerated dramatically. Within a few weeks of being dethroned, these imposing individuals lost weight, lost most of the hair from their massive mantles, had their reddish faces turn dark, and began to look elderly. These changes were not caused by wounds, because they had not been seriously injured, but rather by their loss of status, which correlated with a decrease in testosterone. Mild stress may cause a slight increase in testosterone in high-ranking males, but severe stress always lowers it and, along with it, a male's secondary sexual characteristics (Sapolsky 1990).

With the loss of his females, Padorn spent much of his time with his son Hajo, who had stayed with him rather than going off with his mother. They groomed and defended each other, which they had never done before Padorn's overthrow (Kummer 1995). Then Padorn disappeared, either killed or gone to join another band.

Elders Rosso and Admiral stayed on with the band. Neither tried to obtain new females, nor did either again look at their former spouses, no matter how close they might be sitting. Both spent a great deal of time with their offspring, especially their young sons, hanging out, grooming them, and being groomed in turn. They had become "good fathers," behaving very differently than they had in their glory days as "harem" males. With time, Admiral gained back some of his power, and sometimes acted as a sentinel for the clan, sitting in the hot sun at elevated observation points to spot possible danger. When there was a battle between his band and another, he and the usually timid Rosso took a major part in it, even though the females they were defending were no longer theirs; they were among the few wounded for their efforts. Kummer reported that older females were equally as willing to take risks as older males, exploring and wandering dangerously far from their troop. Elderly males tended to leave the band's sleeping area in the cliffs earlier than the other animals, and, since the other baboons often followed them, they formed a sort of "frontier guard." After half an hour or so of leading, Padorn, for example, retreated to the back of the moving group (Abegglen 1985).

Hans Kummer (1995) revisited his baboon research site in middle age, and one day he decided to try and keep up with the baboons as they foraged. His graduate students suggested he follow old Admiral, which he was able to do, despite Kummer's apprehension that he would now be too slow. The two primate elders walked together steadily for hours, with Admiral a few steps ahead. "At every erosion gully, he knew an easy way to cross, so I never again had to pull myself up vertical walls by clinging to roots." Because Admiral seemed to know in advance where the clan was going, he and Kummer never had to take detours. Admiral also knew about possibly life-saving features en route. Once, when his clan was thirsty, he veered off their path and climbed a bare hill to a dense patch of shrubbery. He entered a passage between two bushes, where Kummer later found a hole in the granite, partly full of water from earlier rains. When Admiral reappeared, the other baboons—including the leading females, who had not known about this well—pushed into the gap for a drink, too.

Kummer's book summarized the life of older hamadryas baboons in general. Some males wander unhappily from one troop to another after they have lost their families, but others fight harder for their bands than even the young males do, face greater danger, and accept more discomfort. Aging females and males are still involved in the life of the clan, and have merely retired from the compulsion to succeed in reproductive life.

Olive Baboons

Since the start of Kummer's field research in 1960, many hundreds of biologists have spent tens of thousands of hours watching what baboons are up to in the wild (Smuts 1985), but none pursued a more difficult research question than the American Robert Sapolsky. For his graduate work, this future neuroscientist decided to study the relationship between animals' hormones and their stress-related behavior. Like many other primatologists, he chose savanna baboons as "his" animal for several reasons: they are a social species, active in the daytime, and large enough to be seen at a distance; they live in open areas, where it is possible to watch interactions between individuals; and once they get used to people who do not harass them, they are not shy.

In 1978, as he related in his wonderful book *A Primate's Memoir* (2001), Sapolsky began a two-year stint of work in a game park in Kenya. Other graduate students had studied his troop of about 60 olive baboons before him, so he did not have to habituate the members to a human presence.[2] They soon ignored him even if he was nearby, so he could easily become acquainted with the various members of the troop. He saw which individuals were dominant to others, shoving them out of the way to obtain a choice bit of meat or a resting place; which individuals gave way to all the others; and how adolescents were treated. But what about older animals, our subject here?

To carry out his research, each day Sapolsky settled down near the troop, so that he could dart every male in turn with a blowgun, anesthetize him, and carry him away for a baseline examination. (He had decided not to involve females, whose hormones were in flux from pregnancy or from nursing.) He had to make sure that the darting was done at the same time each day, to control for daily fluctuations in blood hormones. Sapolsky also had to ensure that his target was not sick, had not just been in a fight or mating, was not injured, and did not know that the dart was coming. He also had to wait until the animal was at the edge of the troop, so neighbors would not be aware of what Sapolsky was up to.

Several minutes later, when his target baboon had fallen over unconscious from the anesthetic in the dart, Sapolsky covered him with a burlap bag and took a quick blood sample from his tail. Then he lifted the hidden body—up to 32 kg (70 lb) in weight—and tiptoed nonchalantly away from the troop, trying to cover up any sound of snoring, knowing that if any of the baboons suspected what he was doing, he would immediately be attacked. He lugged the animal to his vehicle, up to half a kilometer away, where he carried out the rest of the examination: blood pressure, cholesterol levels, rate of wound healing, and levels of stress hormones. Then, covered with sweat, ticks, and baboon hair, he lugged the animal back to the troop and sat near him while he regained consciousness, to be sure that he was all right.

In his research over the years, Sapolsky found that although rank was important in baboon society, it did not account for all baboon activity (1994). Independent of their rank, the older or younger individuals who were most involved with the others, frequently grooming and resting beside them, had the lowest stress hormone levels. In contrast, those who were type A—for example, fighting with other males at the least provocation—had, on average, twice the resting stress levels of their calmer troop-mates.

Needless to say, Sapolsky came to know the members of his baboon troop intimately (2001). He knew who was probably related to whom, who was elderly, and how these older baboons had behaved in their prime. Knowing how important social relationships are within a troop, he found, to his surprise, that some older males left their troop to join up with another, well after their glory years were over. Not that they had not emigrated before, since young male baboons normally leave their natal troop and become members of a different one. In all social species, one sex or the other tends to relocate, thus preventing inbreeding. Once the emigrating young males were comfortably settled into their new troop (which took some time, given that all the females were closely related to each other, and the males would be competitors for food to some extent), they often lived there for many years, eventually becoming highly dominant, exchanging daily grooming with some individuals, fighting with others, fathering young, and traveling with the group every day to find food. This was their home.

It seems insane that an older male would leave the troop where he had passed all his adult years. A lone baboon on the savanna is in grave danger from predators. When an adolescent transfers from one troop to another, his risk of death increases up to 10 times. When an aging male transfers again, he is even more vulnerable than before, because he is now slower, his senses are less aware of his surroundings, and he has no compatriots to warn him of danger. Why would he do this, Sapolsky asked himself?

There are several possibilities—biologists love to conjure up hypotheses to cover such conundrums. Perhaps an older male wants to return to the troop of his youth, where his mother bore and raised him, and his sisters and nieces still live? They might welcome and look after him, now that his muscles are stiff and his joints are riddled with arthritis; by now he is lagging behind the group as it forages, and struggling to clamber into a tree when they settle down for the night.

Or perhaps he emigrates because his daughters are now mature and, following the blind reason of evolution, should not mate with him because of possible inbreeding? (Among the sifakas, it is actually the reproductive-age females who drive out their elderly fathers [Sapolsky 1997].)

Sapolsky proposed a third hypothesis, which he supported with well-documented research (1996). Dominance interactions are frequent among male adults—an alpha male forcing a beta male or other subordinate to give up a scrap of food, jostling him aside at a waterhole, pushing him out of a shady resting spot, interrupting his grooming with a favored female, just as a beta male does to males who are subordinate to him. These males do not bother much with the males at the bottom of the pecking order, who are hardly worth the effort—most are puny adolescents, relatively new to the group. But the dominant males badger them enough. When one of these alpha males grows older and is unable to defend himself, the new crop of prime males makes it a point to persecute him. They remember. When Sapolsky compared aggressive acts against these elder males with those against males of similar age who had recently transferred into the troop, he found that the former suffered over twice as many indignities. Both kinds of males were utterly subordinate, crumpled at the bottom of the hierarchical ladder, but the prime males knew who were anonymous strangers and who were their old bêtes noires. What goes around comes around.

Sapolsky wondered if the elders who had been especially brutal to the young would be picked on more than other oldsters, but this did not seem to be the case. The new prime males might not remember exactly what awful things each older baboon had done to them in the past. What mattered was the fact that he was an elder under whom they had suffered when that baboon was in his aggressive prime.

Solomon is an example of an older male who left his home troop. (Sapolsky liked to give his animals Old Testament names.) He had been the alpha male for three years when Sapolsky began his baboon research. Solomon was a ferocious, canny fighter. He had terrorized his troop-mates to such an extent that he was able, just by glancing at an upstart, or at most swatting him, to keep everyone in line. He even knocked Sapolsky off a rock once, shattering his binoculars, so the researcher was terrified of him as well. Although Solomon was getting on in years, he intimidated the others to such an extent that he had not had a major fight in a year.

But then Uriah, a strong young blowhard, joined the troop. Uriah loved to fight, first one male and then another. He seemed too dim to know that Solomon was the (aging) king, so day after day he attacked him, too. Although he was sometimes slashed by Solomon's canines, he came back for more, until Solomon was worn out. One day, as they lunged at each other, Solomon backed up rather than advanced. Soon a high-ranking male, who would not have dared to attack Solomon several months earlier, now did so. The end was in sight. One morning Solomon walked quietly up to Uriah, turned around, and put his head to the ground so that his hind end stuck up under Uriah's nose in a gesture of submission. His reign was over.

In theory, Solomon might have become the second-dominant beta male, but his troop-mates thought otherwise. Once they realized how vulnerable he was, Solomon became fair game for the other adult males. In short order, he was beaten up by eight of them. He became craven and groveling to those with a higher ranking, and violent to those below him. Finally, he left to join a troop to the south, where he would still be a low-ranked animal because of his age, but at least he would be anonymous and not attacked because of his past history. Sapolsky later glimpsed him now and then, when the two troops met and hollered at each other across the river.

Japanese Monkeys

In common baboons, to be an alpha male is to be an ambitious scrapper. If he shows signs of weakness, younger animals attack and try to depose him. In contrast, the fall of an alpha male is far more genteel in Japanese monkey troops. This species has a strict dominance hierarchy that is seldom transgressed, although the leaders are not necessarily the biggest and strongest animals. Pan's story depicts a gradual disassociation from his troop, until he finally lost his position as a leader. He was one of six leaders in the Takasakiyama troop in Japan, which in 1956 numbered 440 animals (Mizuhara 1964). The job of the leaders was to patrol the central core area of the troop, control the behavior of unruly lesser monkeys, look after infants, give tree-shaking demonstrations to impress minions, and chase away solitary males. During 1956, Pan spent more and more time not patrolling and controlling like the other leaders, but hanging out on the periphery of the troop. He was a rough and vicious character, but still active in the paternal care of juveniles. Some mornings he came to the feeding grounds (where researchers set out food for the troop) at the same time as a group of peripheral young males, but they kept a respectful distance from him.

In early 1957, Pan no longer went to the central area as a leader, but remained at the edge of the troop. Those monkeys who did not recognize him at first tried to chase him away, thinking he was one of the many solitary males without status. In late February, Pan allowed the 24th-ranked young male to mount him three times, an admission of subordination. When researchers dropped peanuts between him and another monkey, to test for dominance between the two by seeing who snatched them, he was often shown to be subordinate to the other. In March, when Pan was chased away from the feeding area by a 9-year-old male, he never again went there. He was last seen in April, feeding in a wheat field 300 m away. He had resigned as a leader, rather than being bested in battle. In stepping down, however, he was not accepted as a member with a lower rank, but was instead forced to become a loner, unwelcome in his own troop.

The last days of the Japanese monkey Jupiter depict what can happen when a leader becomes senile and sick, but refuses to give up his mantle of leadership, as Pan did. Jupiter, nearly 30 years old, was the alpha leader of the same troop as Pan, with Titian next in line as the beta male. He had been a top leader for about 10 years, but in the summer of 1960 he became ill, apparently with paralysis of the sciatic nerve. He hobbled continuously, sometimes having to rest on the way to the feeding ground. On December 23, he was absent for the whole day from the central core of the troop, where 20 females habitually clustered around him. The next day, he returned to the core but, unable to move his legs, he had to pull himself along with his arms. The 20 females again surrounded him. Jupiter initiated the troop's migration up into the mountain in the afternoon, but he was left far behind as the troop members swarmed past him.

In early January, Jupiter was still the alpha male, but his authority was slipping. One morning a subleader, Achilles, chased a shrieking estrus female right around Jupiter, something he normally would never have permitted, but now was unable to prevent. For several days and nights before his death, Jupiter hid himself in a bush near the feeding ground. "He never to the end gave up his controlling behavior as a leader or leadership behavior—so far as he could fulfill his duty physically." With Jupiter gone, the transition to Titian as the new alpha male proceeded smoothly; in their long lives, the two males had never shown antagonism toward each other. In 1964, when Titian fell ill and left the troop, Bacchus, the oldest monkey and the male next in line in the ranking hierarchy, became the predominant leader.

Ringtail Lemurs

Ringtail lemurs are of particular interest, because in this species the females are dominant to the males, rather than the usual other way around. "Males never, ever claim rights to food or space," noted Alison Jolly, in her book *Lords and Lemurs* (2004). It is the highly aggressive females who become alpha individuals, and who eventually must lose their high position as they age. Jolly studied lemur behavior in the Berenty area of southern Madagascar, where she described the adventures of Diva, the A-Team's alpha animal and second oldest of her female group. Jessica, her daughter, shared her mother's high status. Beneath them in rank were the fierce ringtail Fish and her 2-year-old daughter Fan. They did not dare attack Diva or Jessica, but they harassed the assortment of subordinates ranked under them who formed the inferior half of the group. Diva seldom had to chase or chastise any of them. They knew enough to keep out of her way.

In the 1992 drought, Diva and her high-ranking colleagues chased the subordinates from the A-Team, so they would have more food for themselves. Because of this, the subordinates, now called the Exiles by the biologists, lacked any feeding territory of their own, and were forced to snatch food from other ringtail territories whenever they could. They were relentlessly pursued by Diva and her high-ranking colleagues; in a series of fierce battles, two infants newly born to subordinate females were killed.

Then the Exiles developed a new tactic—the only innovation Jolly had seen in ringtail lemurs, despite years of observing their behavior. Instead of only a few animals from each group doing the fighting while the others watched, as had been the norm, they now presented a united front, all of them ready to bound (literally) into battle. They claimed their first victory after the five adults gathered their forces on the thatched roof of a bungalow, with Shadow leading the attack as she leapt toward the foe. Their opponents, Fish and Fan, were forced to literally back off, because they were afraid of being jumped on from behind if they turned to flee. The Exiles, in celebration, rushed into a flowering eucalyptus grove from which they had been excluded for a month, ever since the split of the A-Team. There they gorged themselves silly. They had lost two babies in battle, but Shadow still had her tiny infant clinging to her belly upside down.

The Exiles, now called the Together Troop by their human voyeurs, went from victory to victory, claiming eucalyptus tree by eucalyptus tree in the grove, and then eventually—best of all—the former A-team territory. Four males joined the successful troop, something they had been unwilling to attempt as subordinates until they were sure they would survive, individually and within the group.

Diva had made a huge miscalculation when she and the others had expelled the subordinate females from the A-Team. She plummeted from power, battered, worn, skulking on the Together Troop's periphery until she died two years later, a diva or prima donna no more. Jessica, who had been "slender and lovely" when her mother was alive to protect her, became thin and stringy after Diva died. She and what was left of the A-Team never again gained a range of their own. Over the next few years they all died or disappeared, a sad ending to Diva's regime.

Lions

Among lion prides, alpha males fail not individually, but as a group, just as they rule as a group. Jeannette Hanby, in her book *Lions Share* (1982), described what it was like for old Leo on the Serengeti Plains. He was not actually being deposed, but Hanby's portrait depicted him as coming to realize that his aspirations for greater glory were hopeless and that the end of his and his pride-mates' regime was near. This aging male tried to expand their range into that of the Sametu pride, a pride he had often harassed in the past. Leo and his cohorts were in effect ranchers, "owning" a large piece of land and all the prey species that lived on it. This has shades of human capitalism—the bigger the pride territory, the richer and more successful the pride males. Most lions in Africa are killed by poison, snares, guns, disease, accidents, or other lions, but a few, such as Leo and his companions, survive to old age—males into their teens, and females into their twenties.

One night, hearing growls and the noise of meat being sheared from bone, but no roar of males, 11-year-old Leo ventured into the territory of the Sametu females. They were feeding at a zebra kill, while their two males politely waited nearby for their turn at dinner. The females recognized Leo in the moonlight, and immediately got even with him for all the grief he had caused them in the past. They rushed at him in a body, growling, hissing, and clawing, and knocked him off his feet. He struggled up and fled down the hill, the way he had come.

The two males now gave chase, roaring as they ran. Rather than retreat farther into his own territory, though, Leo foolishly chose to turn and face his pursuers. He howled and slashed at them while the young males attacked again and again, biting him and raking him with their claws. Finally, they left Leo crumpled on the ground, choking and gasping for air. The two victors roared in triumph as they returned to the females. Leo staggered away, and was met by his battle-scarred pride-mate Lerai, come to see what the fuss was about. While Lerai continued on to the scene of the conflict to sniff out what had happened, Leo lay down and rested for a long time before lifting his head and beginning to lick his wounds. Later, the two aging males walked slowly toward the center of their territory, soon joined by a third elderly male; there was no way now that they would ever take over the Sametu pride. Among them, they and the two other pride males had fathered a prodigious number of cubs, but before long they would inevitably lose their females and pride territory to males in their prime.

IN SUMMARY, SOCIAL MAMMALS usually have an alpha animal or leader for their group, but when this individual becomes too old to carry on, his or her leadership is terminated in a variety of ways. This does not necessarily mean that the fallen individual is no longer useful to the group, though. In the case of baboons, older males sometimes act as sentinels for their group, lead the group members in the direction they will go to forage for the day, and fight for their troop against members of other troops, even though they no longer have females with whom to mate. The next chapter considers the fall of titans living in captive groups.

CHAPTER 9

Aging of Captive Alphas

THE ONLY PLACE TO VIEW, in continuing detail, the social relationships involving the decline of an aging alpha animal is in captivity, where groups of animals can be observed every day, all day long. Zoologist Frans de Waal undertook such research at the large chimpanzee colony in the Netherland's Arnhem Zoo (1996). He found that chimpanzees living in captivity under semi-natural conditions behave not only like wild chimps, but even more so. Individuals are forced to live close together, so interactions are usually intensified. In captivity, adult females become more social (in the wild they are rather solitary) and, like permanent groups of many monkeys and apes, they band together to protect themselves against male violence, influence male power struggles, regularly groom each other, and share food. Captive adult males, however, actively display both bonding and rivalry behaviors.

In 1971, the Arnhem Zoo decided to form a chimpanzee troop by adding an assemblage of various males and females to a large enclosure. The large, strong, older female, Mama, became the group leader, even though she was then nearly 40 years old (de Waal 1982). The zoo authorities were upset about this, because they supposed that males should be dominant (although in the wild, dominance is not an important concept for chimpanzees—natural food is not limited, and lusty females solicit many males for mating). They introduced three adult males, one of whom was called Yeroen, into the group to rectify this situation. To the zoo men's consternation, these males were no match for old Mama and her female friend, the oddly-named Gorilla. The three males huddled on top of each other on a tall drum to try and fend off Mama and her cohorts, who bit their feet and pulled their hair. The terrified males screamed, vomited, and defecated in fear—a pathetic display for the watching men. After two weeks of mayhem, injury, and tension in the enclosure, the staff removed Mama and Gorilla so that the males could at last gain ascendancy.

Three months later, when Yeroen was the newly acknowledged alpha animal, Mama and Gorilla were allowed to rejoin the troop. Again the entire group erupted in chaos, as it had three months earlier, with all the apes screaming and the three males huddled on top of the drum. When a splatter of feces from one of the terrified males fell onto Mama's leg, she calmly cleaned it off with a piece of frayed rope. But this time the other females did not support Mama in her aggression against the males. Even her buddy, Gorilla, soon made friendly contact with Yeroen. During the period she and Mama had been banished from the group, the pair's female alliance with them had been broken. Soon Mama had to concede defeat as leader, although she remained the diplomatic alpha female and the "mother" of all the chimpanzees (de Waal 1989).

After this, Mama became less aggressive as she aged, more often spending her energy in breaking up squabbles than in causing them. She did not become a wimp, though. De Waal (1982) described how she was defeated by (female) Oor when Oor persuaded (male) Nikkie to join an attack on Mama on Oor's behalf. When Oor and Mama were alone together after lockup, however, Mama thoroughly trounced her rival. Mama refused to nurse her next young, born after her defeat as leader, but even though she was aging, she became an ideal mother to the infant born after that. Mama lived on for some years. De Waal (1996) reported that every time he returned to the Arnhem Zoo, she recognized him, even when he was among hundreds of other visitors, and would "move her arthritic bones to meet me at the moat's edge with pant-grunts."

As Yeroen, too, aged, having become the alpha and oldest male of the Arnhem troop, he began to express a crafty side of his nature (de Waal 1996). When he hurt his hand in battle with an up-and-coming younger male, even though the wound did not look serious, he began to feign infirmity, but not in his hand. He would walk normally until he was aware that the other male could see him, and then begin to limp. Was he hoping that his apparent injuries would elicit sympathy, rather than aggression, from the other male? Was Yeroen able to imagine how he would look to his rival as he hobbled along? Or had he by chance been limping on another occasion, and found that this was a suitable way to turn away wrath?

The first few times that Yeroen was bested by younger males—especially by Luit, who would replace him in the alpha role—Yeroen dropped his normally dignified mien. He became so upset that he threw temper tantrums (de Waal 1996). He would scream and beg with his arms out for a female, someone, anyone, to help him, making such a racket that de Waal could hear him a kilometer away. Sometimes an older female or a youngster would go to him, and put an arm around his shoulder to try to soothe him. Occasionally, he would fall out of a tree, writhe on the ground, and scream to attract attention, acting like a juvenile no longer allowed to suckle from his mother and, also like a juvenile, keeping track of who was willing to give him support. If there were enough chimpanzees crowded around him, he would gain instant courage, and, leading these sympathizers, rekindle the fight with his rival (de Waal 2005).

After a time, Yeroen became more reasonable. He had two options: to support Luit, the new alpha male, and obtain a few benefits because of this, benefits determined by Luit; or back another male, who would then owe this top position to the support Yeroen gave him (de Waal 2005). He decided that rather than fight losing battles in trying to retain his own position as alpha male, he would back young Nikkie as the alpha chimpanzee in the group. Nikkie was inexperienced, and heavy-handed in his interactions with others, so he could learn a great deal from Yeroen, who was impartial and controlled while mediating fights (de Waal 1996). The other animals trusted Yeroen to be fair; together, the two chimps made a good team.

Eventually, though, Nikkie became more confident and bossy. Sick of deferring to his elder, he took to breaking up Yeroen's mating attempts, to the latter's rage (de Waal 1996). Then Nikkie, knowing that he would no longer be top chimp without Yeroen's backing, would quickly hold a hand of friendship out to him again. The aging chimp Mama sometimes facilitated such a rapprochement. On one occasion, she went to Nikkie, who was screaming, and put her finger in his mouth as a gesture of reassurance. Then she nodded to Yeroen, and held out her hand to him. When he came over to kiss her on the mouth, she stepped back from between the males so they could make up their quarrel, which they did. Together, they began to chase the rival male, Luit (de Waal 1989).

After three years, the alliance between Nikkie and Yeroen fell completely apart, and Luit became the alpha animal. How would an older male such as Yeroen go about acknowledging Luit as his superior? To find out, researchers carefully documented all interactions between the two males for three months (de Waal 1996). Former friendly contact between the pair had quickly disappeared, while intimidation displays increased. Soon Luit began to turn away when Yeroen approached, upsetting him. Finally, Yeroen uttered his first submissive pant-grunt. Immediately, their relationship underwent a radical change. Luit made friendly gestures to Yeroen, at which the other chimps, watching the two, immediately rushed over to embrace both of them (de Waal 1989). Within a few hours, the pair had settled down to groom each other. Because Yeroen had acknowledged that he was no longer dominant, the duo could now become friends.

This friendship, though, did not last. One night when Yeroen, Nikkie, and Luit were caged together, a terrific fight broke out when there were no keepers present to stop it (de Waal 1989). In the morning, they found blood smeared over the walls and floor, and Luit mortally injured. Nikkie and Yeroen must have collaborated together once again to destroy him, and crafty old Yeroen would continue to be involved politically in the group's leadership.

At the scene of the catastrophe, de Waal bent down to Luit, who was sitting in a pool of his own blood, leaning against the bars of the night cage. The primatologist gently stroked him and Luit, who had never before been affectionate with people, seemed glad of this comfort. De Waal (2005) wrote that this seemed to him to be an allegory of modern humanity: "Like violent apes, covered in our own blood, we long for reassurance. Despite our tendency to maim and kill, we want to hear that everything will be all right." De Waal rushed Luit into surgery, where the physician sewed up his many wounds with hundreds of stitches. Luit had lost fingers and toes, and even his testicles had been squeezed out of his scrotal sac. His case was hopeless, however. Luit was so badly wounded that he died while under the anesthetic.

Francis Fukuyama, a renowned professor researching public policy and international affairs, cited this battle among the three male chimpanzees to show that aggression in men is an inborn trait (1998). He insisted simplistically that if nations are to survive, masculine policies of aggression, rather than feminine tactics of cooperation, will always be essential in the world of foreign affairs.

Before Hans Kummer carried out his research on wild hamadryas baboons in Ethiopia, he studied members of this species in 1955, in the Zurich

Zoo in Switzerland (1995). The animals lived in a large enclosure, where there was plenty of room for them to act out typical baboon behavior. One of the two adult males, old Pasha, had four adult females and two young females in his "harem," while the other younger, less-dominant male, Ulysses, had a young female and a number of young juveniles.

By 1958, three years later, Pasha was losing dominance to Ulysses (who was now as large as Pasha), and in some cases to Kalos, a third adult male. Pasha moved from one area to another more frequently than before, always checking over his shoulder to see that his retinue was following him. By April, Pasha had lost much of the hair from his mantle (a condition reflecting a loss of the male hormone testosterone), so that his body looked thin, rather than robust, and his head unusually large. In the evenings, he waited until last to be fed, which had not happened before. At night, he slept with his head on the neck of one of his females, rather than erect, as it used to be. By December, he had lost his two lowest-ranked females to the two other males. One of these females was the wizened, elderly Vecchia, who had joined Ulysses.

With the loss of part of his "harem," Pasha's behavior changed. As the alpha male, he had had little time for the young baboons, but now he lip-smacked, and was content to have them frolic around him. They became his companions. The young male Glumo now groomed the few bristles left in Pasha's mantle, and when the young males rough-housed, Pasha occasionally thrust his toothless jaws into the wrestling group. Sometimes he even initiated a game. Kummer reported that his playing was not a senile reversion to childlike behavior, because he continued to lead his two females about energetically, but rather a choice on his part—as if now that he was no longer interested in reproduction, he was open to pastimes such as those of a bachelor. Despite his shrunken appearance, he still intervened in group disputes.

Six months later, Pasha's high ranking "wife" Sora abandoned him, leaving him only young Liba, with whom he became most indulgent. If Liba did not follow him, he went back to wait for her. He groomed her more often than he had ever previously groomed a female, as if to keep her attention.

Strangely, when he had lost all his females and was but a skinny elder, Pasha attacked the zookeeper for the first time. Kummer suggested that this assault had an evolutionary basis; since he had nothing left to lose, he was free to help protect the other baboons, who carried some of his genes. Or maybe he was just upset because he had no more females to boss around.

The change in Vecchia—from a low-ranked older female traipsing after Pasha to the senior "wife" of Ulysses—was electrifying. Not that the sex was great for her. Ulysses was so thrilled to at last be able to copulate undisturbed (something he had never known before, because Pasha would not tolerate such activity), that he often bit her in erotic excitement, to which she responded by crying out and grooming him. Many months passed before he could have sex with more equanimity.

In other spheres of life, though, Vecchia's personality blossomed, because Ulysses was not really cut out for the job of an alpha male. When Ofe, one of Pasha's former females, made advances to Ulysses, Vecchia charged at her (a female who had once been her superior), bit her on the neck, and shook her, as only a very irate "harem" male would normally do. When Ofe fled to Ulysses for comfort, he ran away. From then on, if Ulysses gave a neck bite to an errant female, Vecchia often gave her a bite, too, for good measure. Eventually, the aging Vecchia took over all the functions of an alpha male (except for sexual intercourse), rising from a cringing, low-ranking "harem" female to the executive officer of the family, despite her sex and advanced age. When Ulysses copulated with Sora, something Vecchia (as a female) could not do, she involved herself by biting Sora on the leg. When Sora went behind Ulysses to groom him after the copulation, Vecchia was stationed there to demand that she herself be groomed instead. Kummer concluded that Vecchia was not superior to Ulysses himself, but rather learned to do what he, Ulysses, left undone. This same type of behavior occurs when baboons are kept in all-female groups in zoos, with the alpha female behaving and being treated in most respects like an alpha male.

PRIMATES BEST EXEMPLIFY THOSE losing their dominance with increasing age, since they have the most fully developed social lives of all species. Individuals accept their fall from power in various ways. The chimpanzee Mama became less aggressive as she grew older, sometimes even acting as a mediator between warring individuals. Yeroen had full-blown tantrums when he was first bested in battle, after which he became crafty and sly in trying to maintain dominance. He made a strategic decision to back Nikkie in order to defeat Luit as alpha male, and then later to submit to Luit when this coalition broke down. However, his aggressive spirit was still present, since he and Nikkie together later killed Luit. Pasha, the hamadryas baboon, lost not only dominance as he aged, but also his rich mantle of fur because of his correlated drop in testosterone. His females left him, and he became a diplomat who liked to spend time with youngsters. However, he, too, continued to harbor aggression, attacking his keeper for the first time. Ulysses finally became, in theory, the alpha male of his family group, but he allowed his partner Vecchia to take over most of the functions of this status. A novelist could write a bestseller based on Ulysses' life alone. These dramas are less evident in the next chapter, which considers small groups where old males and females are no longer stressed by power and hierarchy, but live together in functional, contented families.

CHAPTER 10

Happy Families

THE TOUR GROUP GAPING OUT of the parked Land Cruiser on the Kenya savanna was mesmerized by a small lion cub stalking a large resting patriarch through the short grass, stopping every meter or so to rise up and check his direction, then flattening himself again on the ground (Ross 2001). The big male turned his head slightly to watch the cub's progress, displaying his yellow, worn teeth and drooping jowls. When the cub was just over a meter away from his prey, he paused, gathered together all his forces, and launched a deadly attack at the throat of the 200 kg (440 lb) male. To the group's astonishment, the huge male was knocked flat by the weight of the 5 kg (11 lb) youngster, groaning loudly and taking the cub with him as he toppled over. The cub clambered triumphantly along his prey and, gaining a grip with his teeth, shook the patriarch's neck as hard as he could. The massive male groaned again, pawed the air briefly, and then lay still. The cub slid to the ground and pranced on to find his next conquest. The male regained his resting stance and again gazed calmly into the distance.

Every few years, pride males do battle with—and are often replaced as pride males by—a younger or a larger group of males, in which case they usually become aging nomads, wandering over the plains of Africa in search of food. But occasionally, elderly males, perhaps 10 years old or so, hold on to their prides and are able to play with their cubs, which is what had happened here.

"Family" groups of animals contain females and, usually, their young, but this chapter discusses those that include males as well, or that are even made up solely of males (who are in general less sociable than females). The individuals discussed come from highly social species—dogs, wolves, baboons, and gorillas. Even though we currently do not know if a male of these or other species recognizes a youngster he has sired, this has proved to be the case for wild chimpanzees (Lehmann et al. 2006).

Dogs

Elizabeth Marshall Thomas decided to carry out a new study on the behavior of dogs. The scores of millions of us who have dogs as companions wonder what new things one can learn about them (Grandin and Johnson 2005), but dogs who live with people, especially as their only pet, behave differently than dogs in packs. Pet dogs often feel desperate to communicate with human beings, their only associates—barking hysterically at the postman, pawing a human knee for attention, frantically licking a human face because they have not, as they feared, been forsaken. In effect, people become members of their pack, so much so that dogs hate to be separated from them. Many stories attest to dogs' loyalty to their people, lasting until the death of one or the other. Thomas's ingenious idea was to find out how dogs behave toward each other in their own pack. In her book *The Hidden Life of Dogs* (1993), she recorded the social interactions among 11 dogs she knew intimately, 10 of whom lived with her as a happy family throughout their entire lives.

It is the dogs' final years that interest us here, when they had set aside their youthful exuberance and faced life with a quiet dignity. By that time, Thomas had become fully attuned to their world. She spent afternoons resting with them on a hilltop near the den they had dug in her backyard—sitting as they did in the dust, or lying as they did, on their chests, resting on their elbows, evenly spaced apart. They watched the trees to see if anything moved. They listened to a leaf as it rustled on the ground. They looked at the sun as it began to set, each happy to be there, quiet, and serene. Thomas wrote that "primates [such as herself] feel pure, flat immobility as boredom, but dogs feel it as peace."

When the last three dogs who lived with Thomas in New Hampshire were very old, she installed a dog door in her house so they could come and go as they pleased. There no longer were rivalries among the animals, who now spent most of their time together. Each morning they slowly trailed outside to urinate. The lowest in the hierarchy (owing both to parentage and to her age), the husky Inookshook, first nosed out a suitable place to do so. When she was finished, she straightened up and moved aside so that Fatima, a spaniel/dingo cross, could use the same spot. Finally the high-ranking Suessi, a white husky, lifted his leg and urinated there, too.

Together they would then wander to the top of their bare hill, where they spent the day lying down, calmly watching any activity below. In the evening, deer might come out of the forest to graze, but the dogs were not interested in hunting. More often they watched the road, in case a neighborhood dog should meander by.

One day, after the Thomases had mowed the field below the hill, a coyote arrived to feed on the small insects and mammals that were exposed. On seeing this interloper, the dogs streamed down the hill toward her. Thomas was horrified, because she feared a battle was imminent. She yelled at the dogs to come back, but they ignored her. She ran after them, thinking she might scare the coyote away, but before she had gone far, the dogs had stopped about 10 m from the newcomer, who had straightened up at their approach. All four of the animals looked at each other; then they all began feeding on tidbits uncovered by the mower. Thomas realized that the dogs and the coyote knew each other, and were friends.

At the end of her book, Thomas declared that what dogs want in life is to be with each other. When they live in a pack, they remain calm and sensible. They understand their fellows, and know their place in their small circle. Inookshook, although the lowest-ranked of Thomas's three dogs, was fully accepted by them, and was accepting of her own place. What human beings were up to was no longer of much interest to them, as long as the humans supplied food, water, and shelter.

Thomas's dogs did not sing, as wolves may do, at the first glimpse of the red sun each dawn on the eastern horizon, but Timothy Findley, in his book *Inside Memory* (1990), reported that his dogs *did* honor the sun. He joined them in their hutch one hot night to share their private life. After he and the male had walked around the periphery of the pen, marking a few spots with urine in case there were marauders, they settled down in the hutch with the male's mother. Findley, in his sleeping bag, was given the place near the screened window, where it was cool. Both dogs were restless throughout the night, dozing rather than sleeping soundly. Near dawn, they looked at each other simultaneously. The mother rose and walked to the door, followed by her son. Findley crawled after them. They all faced the east and, at the exact moment that the edge of the sun breasted the horizon, they bowed, Findley perhaps a second late because he did not know the protocol.

In the process of living their own lives among us, older dogs give courage and support to their human companions. Susan McElroy, in her book *Animals as Teachers and Healers* (1996), related the story of an old greyhound, Echo, who was adopted after the end of her racing career by a woman who had suffered a recent divorce. The woman credited Echo with giving her back her life when she was clinically depressed: "I learned so much about quiet presence, dignity, love, and tolerance from this elderly gentlewoman. I owe her my health and well-being."

Wolves

Male wolves, close relatives of dogs, also have strong social connections. A pack of five wolves in Alaska was observed by scientists over a seven year period. The pack had a large male as the father, easily recognized by his pronounced limp—his left front foot was so swollen and sore that he held it in the air, and rarely put his full weight on it. It had probably been injured in a steel trap, perhaps set by a trapper wanting a bounty or a pelt. The old wolf had memorable grit. When he chased caribou, he put his injured leg to work, which often gave him enough speed to catch his prey, but then he would lie down on the tundra in apparent pain, licking his maimed paw. The other pack members were a white alpha female and three of their progeny, all light gray in color. Wolf expert Rick McIntyre (1993) lyrically described the pack trotting eastward, the alpha male falling behind because he moved only on three legs, while the other wolves repeatedly paused to wait for him. At each stop, the youngsters touched noses with the female or rolled about near her on the ground, all in high spirits. To McIntyre, their behavior demonstrated the sheer joy of belonging together in a pack.

In what was surely his last summer, the alpha male decided to feed his pack by killing an adult moose, weighing five times as much as he did, on a solo hunt. Over 36 hours, he attacked the animal at least 16 times, dashing in to bite the moose while his prey fended him off with kicks. The wolf's paw was re-injured in the conflict, bleeding profusely, but he still persisted. At last the moose waded into the deep water of a swift river. The wolf swam out to him, where they continued to battle, the moose at one point holding his attacker underwater until he nearly drowned. But now the moose was so weak from loss of blood that the wolf was able to finish him off. His pack soon rushed up to help the victor devour the carcass. However, the contest had weakened the alpha wolf, too. His injuries slowed him down for the next few weeks, and after a month, McIntyre never saw him again.

Baboons

There is plenty of fighting in a baboon troop, but there are tranquil times, too. Barbara Smuts (1985) described the Eburru Cliffs troop, living near Gilgil in Kenya. Elders Virgil and Isadora were sitting near each other, surrounded by a playgroup of a dozen juveniles and infants. Isadora's daughter Venus, who had just finished her first sex swelling, was nearby. Venus watched Virgil snooze for a few minutes, then got up and approached him from behind, reaching out to give his large furry body a hug. Startled, Virgil jerked about to find out who had disturbed his nap. When he saw Venus, he nuzzled her neck while she groomed him.

In this age of acknowledged single-sex unions, we should also consider the close friendship of two older male baboons in this troop, Alexander and Boz. Boz, who was the elder, sometimes defended Alexander in battle. Once, he heard Alexander scream from 50 m away, because the latter was being attacked by a young prime male. Boz rushed to the scene and leapt onto the assailant's back, rescuing his friend.

As they woke up each morning, one of the pair would seek out the other for a ritual greeting. These were of three types, all brief, and all practiced at times by all adult males who, with the exception of the elders, otherwise regarded each other negatively, as competitors (Smuts and Watanabe 1990). Each greeting began when one male, usually the dominant partner, A, approached another male, B, with a rapid, swinging gait while staring into B's eyes and often lip-smacking. If B returned the stare, often lip-smacking at the same time, the greeting continued with B either using a hip grasp on A, diddling A's genitals, or mounting A. If either broke off the sequence of events by turning aside, the greeting was aborted.

The researchers who observed the male greetings—637 in all, during 93 hours of observation over four months—found that in a number of ways, elders Alexander and Boz had a closer relationship than other duos of baboons. They paired up to lure estrus females away from younger prime males (Alexander and Boz were each too old to be able to do this on their own); they cooperated in protecting each other from the aggression of high-ranking males; they greeted each other much more often than did most pairs; the greeting involved mounting behavior (seldom seen between prime males) more often than hip-grasping or "diddling"; almost all of their ritual greetings were completed (unlike the greetings between two prime males, which were often broken off); they used genital diddling during greetings, the seemingly most risky and intimate form of contact, more often than other dyads did; and, most strikingly, they were the only dyad that showed complete symmetry in their active and passive roles. If one was the mounter on one occasion, then the other would do the mounting the next time; the same thing occurred with diddling and hip-grasping. It is wonderful to think of these two aging friends interacting so positively and productively in a community where the younger prime males around them were consumed with aggression and combativeness, rather than friendly cooperation.

In his book *A Primate's Memoir* (2001), Robert Sapolsky describes baboons in a Kenyan game park. The calm baboon Joshua was effectively "married" to Ruth. He, unlike so many of his troop-mates, had not eaten tubercular meat left out by a butcher, and so had survived into old age. He had a longtime relationship with the primatologist, having joined the troop a year before Sapolsky did. He had fallen for Ruth, a nervous female ignored by the other males, and, after a checkered history, they produced a child, Obadiah. Unlike most males, Joshua was a devoted father. When Ruth was tired, he would carry the infant about. Joshua helped him climb trees and, probably to Obadiah's annoyance, interfered in the youngster's wrestling bouts with his friends, apparently worried lest he be hurt.

Joshua had briefly been the alpha male of the troop, and had later supported Benjamin as his successor, but he was less aggressive than the other males, and so avoided most of the fighting which obsessed his peers. In his old age, he sat among the infants as they played, greeted the females in a bewildered way, and plodded along, farting at frequent intervals, at the very back of the troop when it went in search of food.

Sapolsky darted him, anxious to obtain blood and hormone samples for his research, but worried about Joshua's frailty. However, all went well, and when Joshua recovered, Sapolsky began to untie the cage door. As he did so, Joshua pushed his arm through the bars so he could place his hand on Sapolsky's foot. When he had wrestled the door open, Joshua did not leap out and tear off, or turn and try to attack his jailer as the other males did, but rather stepped sedately onto the ground and sat down nearby.

Sapolsky and his wife, Lisa, settled down on the ground beside him, so they could all share a few digestive biscuits. Joshua lifted each biscuit delicately to his mouth with his twisted old fingers, and nibbled on it with his toothless gums. Together they sat in the sun, looking at giraffe in the distance.

Gorillas

Dian Fossey (1983) described the family life of two groups of gorillas, each headed by an elderly male. One was led by Beethoven, the mate of feisty Effie (discussed in the next chapter). He was the dominant silver-back of Group 5, and the father of most of its young. At 160 kg (350 lb), he was twice as big as any of the six females of the group. Beethoven had silver fur along his thighs, neck, and shoulders, and his middle back "saddle" was almost white. He gave deeper alarm "wraaghs" than the younger silverback Fossey called Bartok, and the blackback Brahms, both of whom helped Beethoven protect the troop's females and infants. (The adult males of a troop fight to the death if danger from other male gorillas or poachers threatens the young.) The females and their offspring were all anxious to be near Beethoven and groom him, but, not needing to reinforce his status, he seldom bothered to groom any of them in return. Beethoven was gentle with the youngsters, and placid most of the time.

As the leader of Group 5, Beethoven had a mental map of the places where food plants grew, and of the times at which that food was available, such as in the fruiting season (Weber and Vedder 2001). In addition, every few months he led his troop to a narrow cave in a ravine. Each animal in turn would edge into the cave to consume clay, which was high in iron content. As they traveled between feeding spots, he led the troop along specific routes that he had learned on earlier occasions, with helpful features such as a fallen log, bridging a ravine, that the gorillas crossed in single file.

Beethoven was a considerate leader. When one of the troop's elderly females, Idano, fell sick with a bacterial infection, he slowed the travel pace of the group so that she could keep up. The night she died, he had constructed his night nest near hers. (Idano had recently miscarried, perhaps the result of an attack on the group by a lone male, in which an infant had been killed.)

Beethoven also displayed fatherly behavior. When the female Lisa defected to another troop, presumably because Beethoven had not mated with her for over a year, she left behind her 4-year-old son Pablo. The next day, when it poured rain, Pablo hurried to nestle at the side of rain-gear-clad Amy Vedder, one of Fossey's gorilla researchers. (Vedder was in the process of observing gorillas feeding, spending over 2,000 hours at this for her PhD thesis.) In their book *In the Kingdom of Gorillas* (2001), Vedder and her husband Bill Weber reported that soon after Pablo came up to her, he nudged his head under her arm, so that the two sat huddled together. Vedder was sorely conflicted. Should she act as young Pablo's surrogate mother to possibly save his life? Or should she force him to fend for himself in whatever way he could manage?

The scientist side won out. After that, when Pablo approached her to snuggle up, she kept her arms close to her side to resist him. None of the female gorillas was willing to adopt him, but to the researchers' amazement, Beethoven decided to do so. During rainstorms, he allowed Pablo to shelter under his massive body. Father and son often rested side by side during the day, and slept together, warm and secure, in Beethoven's night nest.

Beethoven was still dominant in Troop 5 when he was about 47 years old, but by then he was strongly dependent on his son Icarus for support in battle (Fossey 1983). In one conflict with a neighboring troop, Beethoven's arm was badly hurt. Icarus, then 14, suffered deep bite wounds on his arms and head. For weeks following that battle, Beethoven and Icarus lay resting in their day beds, their heads together, making soft belching noises as if to sympathize with each other's wounds. However, Icarus's injuries healed faster than Beethoven's. Becoming bored with lying around, Icarus began to lead the other members to feed up to 100 m away from Beethoven. Beethoven would sit alone with his head cocked, listening to faint sounds from his troop. Then he would rise stiffly—quite a few elder gorillas, like many older people, suffer from arthritis (Rothschild and Ruhli 2005)—and slowly join the others. By this time several females had left the troop, and Icarus himself had begun to mate, carrying on Beethoven's heritage. Such partnerships between older father and prime son do not occur in one-male families of other species—such as langurs, guenons, and baboons—where the new dominant male replaces the aging male leader (Yamagiwa 1987).

Beethoven was becoming a little touchy as he grew older. On one occasion, young Pablo grabbed Fossey's glove, waved it in the air, and threw it into Beethoven's lap. Startled, Beethoven leapt up with a loud shriek of terror, scattering everyone around him. The gorillas waited tensely for a few minutes to see what was the matter, and then resettled themselves when it appeared that all was well. They gave Beethoven quizzical glances as he also returned to his day nest, pointedly ignoring the discarded glove (Fossey 1983). On another occasion, young Tuck grabbed Fossey's altimeter and whirled it around her head on the end of its string as she ran to Beethoven. He was scared of it, too, screaming at the twirling missile before quickly retreating (Mowat 1988).

Fossey watched another gorilla couple live to a great age, Rafiki (the leader of group 8) and Coco. Going by the appearance of his greatly silvered coat, Fossey judged Rafiki to be about 50 when she first encountered him. The name Rafiki means "friend" in Swahili, and reflects the great esteem she had for him. She wrote that "in all my years of research I never met a silverback so dignified and commanding of respect."

Rafiki was gentle with his aging mate Coco, so named because of her light-colored brown fur. They acted like an old married couple, often sharing the same nest, sitting companionably side by side, and grooming each other. Coco looked older than Beethoven, with her wrinkled face, balding head and rump, gray muzzle, flabby hairless upper arms, and multiple missing teeth. She presented a pathetic figure, often hunched over, her lower lip hanging loose, her eyes running, one arm lying on her chest, and the other rapidly patting her head.

Fossey suspected that Coco's senses of sight and hearing were deteriorating. One day when she was watching the group from a hiding place, she saw Coco wander away from the others while feeding. Rafiki, who was farther up the mountain, gave a grunt. Coco paused, and then turned to climb toward him and the other males, all of whom sat patiently waiting for her. When she came within sight of Rafiki, she approached him directly and they exchanged hugs, looking steadily into each other's eyes.

Then they walked slowly uphill with their arms around each other, murmuring softly together, and disappeared from sight over the top of the hill.

Coco and Rafiki were the parents of young males Samson and Peanuts, their last young. Coco had taken Fossey's sudden, initial appearance into their lives completely for granted, not at all bothered by a human being observing her from 20 m (65 ft) away. Instead, after glancing at this intruder, she settled down next to Peanuts, with her patchy rump almost in his face, so that he would feel compelled to groom her. Two years after Fossey met Peanuts, he became famous worldwide as the first wild gorilla ever to touch a human being, the hand of Dian Fossey. The spot where this happened is now known as "the place of the hands," to commemorate this wonderful event.

Coco was unexcitable in her old age. Once, when Rafiki and the other troop males were interacting noisily with the males of Group 4, making a terrific din with their displaying and hooting bodies flying through the air, Fossey came across Coco huddled against a tree trunk, one hand tapping the top of her head and the other crossed over her chest. She glanced at Fossey and gave a big sigh, as if expressing resignation at the huge male commotion swirling around them.

Rafiki curbed the movement of the group to accommodate Coco as she slowed down, until one day both Coco and Rafiki were missing. Backtracking their movements, Fossey found that they had slept in connecting nests for the previous two nights, but then she lost all trace of them. Two days later, Rafiki returned to the troop alone, so Coco was assumed to have died of old age. Coco had given cohesion to Group 8; with her death, as the last female, the males began to squabble among themselves.

Although well on in years, Rafiki remained the leader of his small group and, three and a half years later, actually managed to persuade two females to join him. Unlike Beethoven, who had been good friends with his son Icarus, Rafiki was intolerant of his son Samson. The latter left to live a solitary life (something unknown to females), before eventually forming his own small troop. Rafiki, invigorated by the females, soon sired an infant, but he no longer picked fights with males from other groups. He did not live to see his daughter Thor grow up. Gradually, he weakened and no longer moved or ate much; his family foraged for food in a circle around him, so that they were never far away. When Rafiki was

55 or 60, he died on the mountain slope, of pneumonia and pleurisy. Because his son Samson had earlier been driven away, there was no male to succeed him, and so Group 8 also perished.

THE FAMILY LIFE OF INDIVIDUALS in their prime can be chaotic, with males fighting for mating access to the females, females squabbling about transgressions within their dominance hierarchy, and infants sometimes snatched or even killed by other groups. When families center on older animals, however, they tend to be much more relaxed and even friendly, with "all passion spent" (the topic of chapter 16).

CHAPTER 11

Mothering—Good and Not So Good

THIS CHAPTER DEALS WITH VETERAN older mothers, some good at their work, and some not so good. In general, older mothers are far more competent than young ones, because the latter lack experience to guide them. Indeed, for orca females, researchers assume that their first infant will die young (Morton 2002). Jeanne Altmann (1980) studied the relationship between mother yellow baboons (*Papio cynocephalus*) and their offspring in Amboseli National Park in Kenya during the 1970s. She found that baboon infants had a better chance of surviving if their mothers were older, and therefore experienced. One sad example was a young female, Vee, whose firstborn was Victoria. Vicky could not latch onto Vee's nipples during her first day of life because her mother carried her upside down and backward, often bumping her head along the ground as she walked. By the next morning, Vicky was weak and dehydrated. The following day she suckled, but Vee mostly continued to ignore her. Vicky remained emaciated and unable to cling properly to her mother for three long weeks, until she died

Good Mothers

Older females are good mothers, if circumstances allow. Many are also important to their group because of their experience; perhaps they function as role models and mentors for younger females. There are no extensive statistical data on how older female animals perform as mothers, but vignettes of the lives of Effie the gorilla, Baggage the spotted hyena, and Number 42 wolf will give an idea of the reproductive proficiency and lifestyle of at least a few individuals.

Dian Fossey studied and lived among mountain gorillas in Rwanda, from 1967 until she was murdered in 1985 by an unknown assailant (Mowat 1988). Effie, an older female named by Fossey, successfully raised at least six youngsters who were sired by Beethoven, the silverback male, and she sometimes replaced him in leading their group. She was tough, with her rank as dominant female elevating the status of her offspring. (Until recently, female gorillas who change groups upon reaching puberty were thought to be egalitarian because of this. However, a recent long-term study indicates that there is a hierarchical ranking, depending on age and group tenure, and that some females can maintain dominance for from 15 to 25 years [M. Robbins et al. 2005].) In any dispute over food, Effie made it clear that she was willing to fight if she did not get her way. When her family group went in search of clay to eat, she was one of the first to enter the clay cave, and she had priority when they feasted on roots (Weber and Vedder 2001). Fossey described her as patient, stable, dependable, and an excellent mother. She endowed her offspring with love and security, which enabled them to become self-confident adults.

The following story illustrates Effie's observant nature. One of Fossey's students was watching Effie and her young daughter, Poppy, and saw Effie suddenly whirl around to face Poppy, who was behind her. The infant had just fallen and was hanging by her neck in the fork of a tree—Effie had become aware of Poppy's plight even before the student had. With a look of horror, Effie rushed to extricate her infant, glancing accusingly at the student as she did so, and was able to save Poppy's life.

Effie's devotion to her own progeny contrasted with her occasional behavior toward other infants. In one instance, she and her daughter Puck seemed to have consumed, and possibly killed, a 6-month-old chimp youngster, judging from bone fragments found in the dung of the two females (Fossey 1983).

Once Fossey tried to integrate Bonne Année, a 3-year-old orphan rescued from poachers, into Effie's Group 5. As the youngster crawled from human arms down a sloping tree trunk toward the group, the gorillas below inspected her closely. At first, Effie's daughter Tuck welcomed the newcomer with a hug, but Effie rushed forward to grab her away. The two tugged on Bonne Année's arms and legs to try and gain possession of her, pulling and biting until Fossey could no longer stand the youngster's cries of pain and rushed to her rescue (something she had promised herself she would never do, for the sake of scientific detachment).

Fossey clambered up the leaning tree trunk with Bonne Année in her arms, but a few minutes later the infant again elected to climb down to the other gorillas, determined to be free. Effie and Tuck attacked her once more, the baby screaming in pain and fear until Beethoven charged forward, roaring, and chased them away. Bonne Année huddled against Beethoven for a few minutes in the rain, but he was not interested in her, and soon abandoned her to the further assaults of Effie, Tuck, and Icarus, Effie's son. Eventually, Fossey was again able to scoop Bonne Année up to safety. The infant was not fatally hurt, and when she was nursed back to health in a few weeks, she was successfully integrated into Group 4. She had a brief fling at freedom before dying of pneumonia a year later.

Effie did not necessarily defer to Beethoven, even though he was the alpha male. During one heavy rainfall, trying to keep dry, she crowded under a huge fallen log with two other gorilla females, as well as Amy Vedder and Vedder's husband Bill Weber. Beethoven arrived shortly, obviously wanting to join the group, too, although there was no room for him. He parked himself in front of Effie, with his chin jutting out defiantly. Rather than give in to him, she gave a series of sharp cough-grunts. Beethoven stood in silence for a minute, then went on to find some other shelter. (The human couple were left wondering nervously what they would have done if Effie had *not* been there to shoo Beethoven away.) Later, Effie would mate with Ziz and Pablo, Beethoven's sons, without Beethoven minding (Mowat 1988).

When Dian Fossey returned to her mountain camp in 1983 after a three-year absence, she was worried that her gorilla friends would not remember her. She followed her Rwandan guide up a winding trail for two hours before she found Effie's troop. She approached the group within 6 m (20 ft), then sat down, and, in her words, "began making Fossey-style introduction noises—a soft series of rumbles like gorillas make when expressing contentment" (quoted in Mowat 1988). Effie glanced her way while chewing a stalk of celery, looked away, and then did a double-take. Effie stared at Dian as if she could not believe what she was seeing. She tossed away the celery and began striding quickly toward Fossey. Meanwhile, her daughter Tuck had also spotted Dian, quickly looked at her again, approached her, gazed into her eyes, and then embraced her. When Effie arrived, she, too, stared at Dian, sniffed her, "then piled up on the two of us and I was really squished. Her and Tuck's plaintive murmurs reached other clan members in the dense foliage nearby, and one by one the other females came over to us." Soon seven of them were entwined in a group hug, resembling "one big, black, furry ball." Effie had remembered her, with obvious affection. Dian wrote that "I could have happily died right then and there and wished for nothing more on earth, simply because they had remembered."

BAGGAGE THE HYENA

Baggage, another good mother, was a spotted hyena (*Crocuta crocuta*), so elderly that her sagging belly brushed the grass tops as she ambled along. She was a meticulous housekeeper, scraping and tidying the entrance to her den whenever she returned from a trip out onto the Ngorongoro Crater where she lived. She was always enveloped in a cloud of dust when she entered the burrow to rejoin her twins, Sauce and Pickle. After her son Grublin was born, Jane Goodall and her husband Hugo van Lawick, a photographer, spent many months watching and filming carnivores living in this part of Tanzania. Goodall wrote down her observations about Baggage in a chapter on spotted hyenas in their book *Innocent Killers* (Goodall in Lawick-Goodall and Lawick-Goodall 1970).

Baggage was a good friend of Mrs. Brown, a less elderly matron who also had young cubs. Often, on a hot day, the two females would wander into nearby reed beds to rest in the oozing mud. They would rouse themselves in the late afternoon, covered in slime, and return to the Den of Golden Grass area where, with short, low-pitched calls, they summoned their cubs outside into the sunlight. Then they would flop down to relax and suckle their young for an hour or more. Not a bad life. Hyena mothers nurse their youngsters for at least 18 months, far longer than other carnivores, because they neither lead the cubs to a kill to eat, nor usually take meat back to them. Baggage did lug the spine of an antelope to the den where her young lived on one occasion; she then rested nearby to make sure other cubs did not share the meager booty.

Of course, life was not always so idyllic. Members of hyena clans are sometimes given to mobbing one of their own. Once while Baggage was peacefully nursing her twins, three females, including the two dominant matriarchs, descended on her, growling and biting her shoulders and back. Baggage crouched down, squealing and grinning in fear, while her cubs screamed angrily because their meal had been interrupted. Baggage slunk off, followed by her twins, but when she settled down farther away to continue nursing, the females rushed her again, forcing her to move on a second time. They did this once more before the small family was left in peace. Jane Goodall had no idea why they harassed Baggage in this way.

Lions are always a problem for hyenas. Once a pride of lions came right up to the den where Baggage and two other mothers were nursing their cubs. The cubs disappeared into their burrow at the females' alarm growls, while the mothers ran anxiously away to watch this invasion from a distance. Luckily, deciding that it would be too much work to try and dig out the cubs, the lions soon passed on.

On another occasion, a marauding lion caught and killed a hyena, Old Gold, near a wildebeest that the clan of hyenas had brought down. None of the predators wanted to eat the hyena carcass at the time, so it rotted in the sun for a day before Mrs. Brown and Baggage tackled it. Baggage's cubs were not fussy, either, joining in the feast after they had first rolled on the decaying corpse.

WOLF NUMBER 42

In 1995, for the first time in the history of the United States, wolves as a species were officially encouraged rather than reviled. That year and the following one, 31 individuals from Canada were released into Yellowstone National Park, in the hope that wolves would again become a part of the natural web of life from which they had, in the past, been harassed, trapped, shot, and poisoned. All of these Canadian wolves were fitted with radio collars, so scientists could trace their movements. Thus we know which animals lived to be elderly, and how each fared at that time.

Wolf Number 42 was a paragon of older motherhood. In a wild rumble after nightfall (described in chapter 12), she and her friends, some of whom, including herself, had been moved from Canada into Yellowstone National Park, managed to attack and kill their alpha pack-mate, the insufferable female Number 40 (Smith and Ferguson 2005). Number 42 showed herself to be as benevolent as Number 40 had been tyrannical. She took full advantage of her victory. Within a few days she carried her pups in her mouth, one by one, to Number 40's den, the traditional denning site of the Druid Peak pack. She not only raised Number 40's orphaned pups there along with her own, but she also welcomed the family of a subordinate third female, Number 106, who had had the temerity to produce young about 3 km away, presumably to escape Number 40's wrath. She also took over Number 40's mate, Number 21, a fine male who had displayed remarkable tolerance of his earlier spouse. In the summer of 2000, three separate litters of pups, 21 animals in all, were raised in the single traditional den.

One night in 2004, near the border of the Druid Peak pack territory, Number 42 met her end during a battle with wolves from a neighboring pack. Number 42 had become so well known to wolf-watchers for her gentle personality and long life that many people mourned her death, some crying on the road near where she had died. The last wolf of the 31 who had been transferred from Canada to Yellowstone was gone.

Older male wolves can also be socially positive. Michael Fox (1980) noticed that one elder acted as a mediator between two younger males from different litters who had never before met. When these two wolves were alone together they fought, but when the "old man" was present, they were friendly to one another and subordinate to him. Fox also noted that the size of wolf packs is about the size of ancestral human hunter-gatherer families. In wolves, population is regulated by having only the alpha pair reproduce; for early humans, it was controlled by other means, such as killing infants and the elderly.

Mothers of Problematic Young

Three other mothers were good at their job in general, but had trouble with one of their later youngsters, although the problems may not have been their fault. These females were Peggy the baboon, Flo the chimpanzee, and Dacca the captive tiger.

PEGGY THE BABOON

Peggy was a favorite of American zoologist Shirley Strum (1987), who carried out a long-term study of the behavior of savanna olive baboons in Kenya. Baboons, along with macaques, have been as important to field studies of behavior as fruit flies have been to genetics. For the past quarter century, many hundreds of scientists have accumulated tens of thousands of hours of observations on their activities, earning numerous graduate degrees and promotions in the process.

Strum knew Peggy for 10 years before her death at age 32, her teeth too worn to have allowed her to survive much longer. Peggy was strong, calm, social, self-assured but not pushy, and forceful but not tyrannical. She had a cataract in her left eye, so she marched along with her head turned to the left to see where she was going. If she wanted to examine anything closely, she held it up to her good right eye. As the highest-ranking female in Strum's Pumphouse Gang, all her young shared this ranking. If one of them got into trouble, Peggy was there to help fight her offspring's battle, if necessary. But Peggy did not fight if she could manage a controversy diplomatically. As she grew older, she gave birth less frequently than she had in the past. Yet Peggy did not lack for company, because her granddaughters tended to crowd around her, anxious to groom her and be groomed.

Although her family was always her first interest and concern, Peggy also had a close, slightly younger friend, Constance; they were almost inseparable. If Strum wanted to find Peggy in the troop, she was usually near Constance, as well as often near Sumner, an older male who frequently hung out with them and their families. If Sumner had a piece of meat, Peggy usually managed to persuade him to share it with her. Sumner made a great jungle gym for Pebbles, Peggy's last infant. Pebbles would leap onto him, clamber up his face, burrow through his thick mantle, and slide down his big belly.

Peggy's life was rich and friendly. The only dark side was her daughter, Thea, who was very different from her mother. Thea was unpredictably aggressive, attacking and harassing others for no apparent reason, and meddling in spats that did not concern her. Thea was, in Strum's words, "a bitch." Most of the troop gave her a wide berth. Thea, being younger, was subordinate to her older mother, until misfortune struck the family. Peggy's son Paul died, baby Pebbles disappeared (and probably died, too), and Peggy became very ill. Thea took advantage of these disasters and seized dominance from Peggy, who was too weak to resist—an upset of the established hierarchical order which is extremely rare in olive baboons. At first, Thea's daughters were confused by this change. Why was their mother harassing their beloved grandmother? But then they began to emulate Thea, persecuting Peggy unmercifully, and even siding with strangers against her, something that was previously unknown. Peggy put up with such attacks with little resistance.

Two years after Peggy died, Thea herself suffered ill health—a painful foot injury which made it difficult for her to walk. With little fuss, she, in turn, was deposed by her eldest daughter (monkey see, monkey do). Perhaps the daughter refused to give way to Thea on one occasion, and because Thea was not strong enough to defend herself, and no one nearby would help her given her bad reputation, her fate was sealed. And deserved.

FLO THE CHIMPANZEE

Another good mother was the chimpanzee Flo, observed for many years by Jane Goodall (1986) in her study area in Gombe, Tanzania. Flo was Goodall's best known and oldest subject, mother of Faben, Fifi, Figan, Flint, and Flame. She became so famous from Goodall's research that when Flo died, she had an obituary in London's *Sunday Times*, which read, in part:

> Flo and her large family have provided a wealth of information about chimpanzee behaviour—infant development, family relationships, aggression, dominance, sex...But this should not be the final word. It is true that her life was worthwhile because it enriched human understanding. But even if no one had studied the chimpanzees at Gombe, Flo's life, rich and full of vigour and love, would still have had a meaning and a significance in the patterns of things.

As Flo grew older, her high social status, based on her feistiness, declined. She was still important to her family, though, and willing to play, either with her own children (grabbing at their feet when they were chasing each other around a tree, for example) or with one of the males. Fifi remained good friends with her mother to the end of Flo's life, grooming and supporting her. However, Flo finally could no longer climb trees to nest or feed, and spent her days in an area so restricted that Fifi had to leave her to find food for herself. Flo was occasionally able to successfully beg meat from a male, but sometimes she was reduced to consuming food remains from feces.

A mother's work is never done. Not long before her death, when her son Figan was 23 years old, he screamed when he hurt his wrist while in a dominance fight with another male. When Flo heard his cry, she rushed half a kilometer to see what was wrong. He was still wailing when she arrived, so she sat him down and groomed him. Thanks to her comfort, his screams gradually subsided and he became calmer. Figan did not really deserve such kindness, because he had attacked Flo several times during the height of banana feeding, giving her a mild pounding. Flo had been enraged rather than frightened at the time, screeching in fury.

Although Flo bore two children when she was older, they did not survive to adulthood. She could not force her son, Flint, to become independent, giving in to his demands to let him sleep with her in her night nest and allowing him to ride on her back like an infant, even when he was over 8 years old. It may be that Flint was mentally retarded, which would explain his unnatural behavior (Jahme 2000). When Flo died, Flint was grief-stricken. He fell into a depression, dying himself three weeks later. Flo's last child was Flame, an infant who disappeared when Flo suffered a bout of severe sickness from which she was fortunate to recover.

DACCA THE TIGER

The tiger Dacca, a lifelong inhabitant of the Bronx Zoo in New York, died at the advanced age of 20, but she had lost her zeal for raising cubs years before. Tigers (*Panthera tigris*) used to be much less common in zoos than lions, so there was a celebration in 1944 when the wild-caught tiger Jenny—who had lived in the zoo for 10 years, together with "a very large but decrepit male"—at last produced three cubs who survived. The keepers were thrilled, even though the cubs had to be hand-reared by staff (Crandall 1966). Dacca was the largest, followed by her brothers Rajpur and Raniganj. Keeper Helen Martini, the wife of the lion-house keeper, undertook to raise them on bottled milk, initially feeding them every three hours, night and day.

Raniganj was removed from their joint cage after three years, because

Dacca disliked and attacked him. She wanted Rajpur as her mate. In 1948, she produced her first cubs. She tended them with great devotion, not bothered when she was interrupted each day by Martini checking and handling her young. Once, when Martini had her arms through the wire front of the cage to greet Dacca, the tiger carried a cub in her mouth, took it over to the woman, and laid it in her hands, imprisoning her in place. As Dacca returned to her den, Martini frantically called to her to come and retrieve the infant, which she calmly did.

Over the years, Dacca produced 32 cubs in 11 litters, of which 28 were fully reared to adulthood with no reported problems from inbreeding. Despite never herself having had her mother's care as a cub, Dacca was a wonderful parent for them. However, her maternal instinct was noticeably waning for her last few litters, all passion spent. The keepers called her final cub Finis, and in 1959 they separated her from Rajpur, so that she could live her last few years in well-earned rest. She remained gentle and friendly with her progeny, and with the Martinis, to the end.

A Grieving Mother

It is not known exactly how good a mother Sylvia was to her daughter Sierra, but certainly she grieved excessively at Sierra's demise. Sylvia was one of a troop of free-ranging chacma baboons (*Papio hamadryas ursinus*), studied by researchers for 14 years in Botswana's Okavango Delta (Engh et al. 2006). When a lion killed Sierra, Sylvia became depressed. She had an increased amount of glucocorticoids in her blood, hormones which are also found in stressed human beings. Stress levels in baboons rise not only with the death of a grooming partner or close relative, but when their own social ranking is at risk.

Sierra had been both Sylvia's daughter and her closest grooming partner. Without her, she was desolate. Sylvia had been known as "the queen of mean," a very high-ranking, 23-year-old monkey who had, until then, disdained female baboons other than her daughter. But with Sierra gone, her need for social contact was so great that she deigned to give up her customary aloofness and groom with females of much lower status. With the help of such friendly connections, her stress hormones soon diminished.

IN SUMMARY, OLDER MOTHERS ARE EXPERIENCED mothers, having raised a number of young during their lifetime who would carry on their genetic inheritance. Some of them have trouble with one of their final offspring, but they carry on as well as they are able. The next chapter will again consider mothers, this time grandmothers, who affect the lives of their grandchildren as well as their own progeny.

CHAPTER 12

Grandmothers

I N THE ANIMAL KINGDOM, males usually become less aggressive when they are past their prime. However, some females become *more* combative as they grow older, as typified by the feisty baboon Vecchia, described in chapter 9. In humans, older men are far less violent than young men, while older women often become more self-assertive, although rarely aggressive. This chapter first considers combative older females: the lives of langur monkeys in India; captive vervet monkeys; two wild Canadian wolves, one of them a grandmother and the other prevented from being one; and prairie dogs. It then examines instances where, according to the grandmother hypothesis, elder females expend energy improving the lives of their grand-offspring, such as sperm whales, langurs, and human grandmothers from 150 years ago.

Aggressive Grandmothers

LANGUR MONKEYS

Langur monkeys living in the vicinity of Mount Abu, Rajasthan, India, lead an exciting if somewhat unusual life, in that they interact closely with people—friendly shoppers who toss them peanuts, chickpeas, or *chappatis* to munch; angry merchants from whom they snatch and wolf down delicacies; and watchful men who patrol garden crops tempting to both humans and monkeys. As Sarah Blaffer Hrdy described in her book *The Langurs of Abu* (1977), their tameness allowed her to observe them, and to make notes on their behavior from 3 m (10 ft) away without alarming them. She watched them over a period of five years, from 1971 to 1975, and found that the age of individual langurs was central to the way their society was organized.

As is usual with all wild animals, it is difficult to determine the exact age of any one adult, although elderly langurs have eyes that are more deeply ringed with creases. Female langurs, in general, continue to grow after their first young is born, reaching full size by the time they are middle-aged—about 12 years old and 11 kg (24 lb) in weight. At this time, they have wider backs and wrinkled, sunken faces. The oldest females have flattened faces, wide nostrils, and bushy manes of silver hair around their faces. They are bony and appear decrepit (misleading, as chapter 13 demonstrates), with various scars and rips in their ears, accumulated particularly in their fighting grandmother years. By middle age, the males have also reached their full body size—18 kg (40 lb)—and are stockier than their younger selves.

A langur troop is usually made up of a number of females and their young, along with an alpha male. The prime adult females are socially active among themselves during the day—grooming, hugging, or mounting each other, as well as feeding, resting, and caring for their young. They are the dominant group. As young females grow, they work their way up in the hierarchy to join this exalted circle, while older ones slowly lose rank as they produce fewer young with age. In Hrdy's research, prime, middle-aged females produced 0.38 young per year, while the four oldsters produced fewer than half as many. A dominance hierarchy reflective of age in langurs is different from that for baboons and Japanese monkeys, species in which the females inherit their rank from their mother and retain it throughout their lifetimes.

Dominance affects a huge number of interactions in the daily life of langurs. The bossy prime mothers grab food from an older female if they feel like it, which they often do, or demand that elders move away from a coveted shady spot. If elderly Circle Eyes tried to feed alongside the others, she was likely to be attacked by younger females who demanded her place on a garbage heap, or at the tasty oozing sap of a palm tree. Sometimes their attacks on her drew blood.

Needless to say, because of this antagonism by the prime females—usually their own daughters, since it is the males rather than the females who change troops on reaching puberty—the older females tended to keep to themselves. Circle Eyes never approached her troop-mates to initiate any social activity, but avoided them whenever possible. She went off by herself to feed, remaining among her troop-mates only when they were snoozing or absorbed in grooming. The elderly female Sol, named because of her solitary disposition, also socialized little with others (although she was eager to fight in their defense), and virtually never traveled with her troop during the daytime.

Being denied a social life within the troop makes older females seem feeble and pathetic, but nothing could be further from the truth. They are incredibly feisty. Despite their elderly appearance and low ranking, older females are the chief defenders of the troop—not only from human beings, but from dogs, members of rival troops who sometimes try to kidnap babies, and alien langur males. Aging females are the ones most likely to threaten, charge, or slap at such enemies (as Hrdy discovered when Circle Eyes often threatened her for being too close).

For example, on August 12, 1972, Mug (a Hillside troop male) was intent on grabbing Scratch, Itch's baby. He sat for a while on a rooftop looking off into the distance, then suddenly rushed down into the tree where Itch was feeding and tried to snatch her infant, nicking the baby in the process. Immediately Sol, Pawless (almost as old as Sol), and an unidentified female counterattacked. They lunged at Mug, forcing him into a quick retreat. Itch was left alone, clutching her blood-spattered baby.

During the rest of August and into September, Mug kept up his harassment of Itch and Scratch, but on all eight occasions, Sol successfully defended the mother and her infant, helped on seven of them by Pawless. On September 9, Itch managed somehow to drop Scratch from the jacaranda tree in which they were perched. Mug immediately raced for the baby, but so did Sol and Pawless, who were nearby. By jointly attacking him, the two older females were able to retrieve Scratch from Mug and chase him away. Scratch managed to survive this attack, but later disappeared and was presumed dead. Older females, despite defending infants, in general were not interested in the babies themselves, and took little or no part in holding or caring for them. Sol, the oldest of all the langurs, disappeared (and may have died) in 1974.

Pawless, who was classed as middle-aged to elderly, lacked a left forearm and hand, but this did not prevent her from bearing and raising young, or from defending her troop against enemies. Indeed Pawless, on one occasion, led an attack against the neighboring Bazaar troop. She charged a watchman intent on driving away her troop, which was feeding in a garden he was hired to protect, and also chased a small boy who hit her with a stone.

On another occasion, two older females from the Bazaar troop, Short and Quebrado, helped abet a kidnapping. A School troop female with a newborn had her infant wrestled from her by a subadult Bazaar member, Junebug. Junebug ran off on two legs, clutching the baby to her chest, pursued by the mother. But Short and Quebrado soon intercepted the mother and forced her to return to her own group. The baby was later passed around among the Bazaar females, but was somehow returned to her mother by the following morning. On another occasion, Quebrado fought so hard against a male attacking an infant that she was knocked out of the tree in which they were wrestling.

Short was equally aggressive. When Hrdy fostered an encounter between two different troops by placing a pile of chickpeas between them, Short led an advance against the members of the other troop, and almost single-handedly drove them back in order to capture the prize. However, far from being rewarded for her efforts, the young alpha female Elfin took possession of the food and gave Short none.

These offensives by old females for the good of the troop are of special interest, because they do not benefit the oldsters themselves, and indeed often leave them bloodied and wounded. Rather, their aggression presumably benefits their offspring in the troop (including their ungrateful daughters), who share some of their genes.

Older females may be pushed aside or ignored much of the time by younger troop members, and allowed to do most of the fighting on their behalf, but they are respected for their life-long experience. They are the most likely members to know which trees in the area are fruiting at any one time, what gardens have intolerant watchmen, where water is available, and which urban areas are most anti-monkey. Hrdy found that in almost every instance when a female decided in which direction a troop would travel, it was an older rather than a younger one who showed the way, despite her low dominance ranking that made the troop largely shun her.

VERVET MONKEYS

The functioning of individuals in wild langur monkey society may be elucidated in part by the findings of a study of 21 adult female vervet monkeys (*Cercopithecus aethiops sabaeus*) living in a primate laboratory in California (Fairbanks and McGuire 1986). Here, also, grandmothers were aggressive in defense of their own daughters, and their belligerence could be shown to have paid off. Young adult females with mothers had significantly more offspring who survived than did their peers who lacked mothers. These benighted individuals were attacked more often by other monkeys, were less likely to be defended from such attacks, and were unlikely to challenge the rank of other monkeys. Grandmothers were therefore important in bettering their daughters' social and reproductive success.

WOLVES

As with monkeys, aggression can play out among female wolves in the same pack. Number 40 and Number 42 were among the last survivors of the original Canadian wolves introduced into Yellowstone National Park. Number 42—not actually a grandmother, but the age of a grandmother— was a member of the large Druid Peak pack, along with Number 40. These females were anything but sisterly in their behavior. In their book *Decade of the Wolf: Returning the Wild to Yellowstone* (2005), Douglas Smith and Gary Ferguson noted that Number 40 was the undisputed alpha matriarch from 1997 to 2000, the fiercest wolf of the entire Canadian group: "one had only to look cross-eyed at this alpha to find herself slammed to the ground with a bared set of canines poised above her neck." Much of her ill temper was directed at Number 42, who was so abused by her cruel sister that in two National Geographic films she was dubbed "Cinderella." In 1999, for example, when Number 42 began digging her own den apart from that of Number 40, as if to have her own cubs there, Number 40 gave her a terrific thrashing. As usual, Number 42 did not defend herself, but merely lay on her back and accepted her punishment. She never again went near this den.

In 2000, Number 42 tried again to have her own family, digging another den and this time producing pups. When they were about six weeks old, she, the pups, and the faithful cadre of younger females who helped her supply food for the pups went for a stroll, and happened upon Number 40, "the wolf avoided by most and feared by all." Number 40 immediately set upon Number 42, and gave her a trouncing even worse than those she had received before. Then she turned on one of Number 42's helpers. At this point, all the wolves were heading toward Number 42's den, but as night was falling, the human voyeurs were left in the dark (double meaning) about what happened next.

In the morning, they found Number 40 so badly injured that they thought she had been hit by a car and dragged along underneath it. After she expired, it was clear that she had been killed by pack members. From radio evidence, it seemed that near her den, where she was on home territory, Number 42 turned and attacked Number 40 after years of abuse, desperate to protect her pups and aided by her female supporters. Payback time. Number 40 was always so bad tempered that she did not have any female followers. She was killed by her own subordinates, the first time this had been known to happen.

PRAIRIE DOGS

Older females are not necessarily aggressors, though, as John Hoogland (1995) found in his study of black-tailed prairie dogs (*Cynomys ludovicianus*), which comprised 73,000 person-hours of observation in South Dakota over 16 years. Infanticide is all too common among this species. The killers were usually lactating females who attacked the infants of close relatives, affecting 22 percent of the litters. This was a shocking revelation, but it is pleasant to report that of the 65 killers, none was age 7 or 8 (the oldest females), and only three were 6 years old. The oldest female prairie dogs may or may not be caring toward their descendants, but they were not killing the competition.

Caring Grandmothers

The grandmother hypothesis postulates that older females, especially those who are no longer reproducing, may use their energy and resources to improve the lot of their descendants, thereby increasing the dissemination of their own genes.[1] One example is sperm whales, who are so difficult to study that we can only assume what is going on. Secondly, langur grandmothers do not spend all their time fighting, and can be solicitous of their grandchildren. Japanese monkey grandmothers also appear to have a positive influence on the survival of their grandchildren, but there may be few daughters who are reproducing who actually have post-reproductive grandmothers available (Pavelka et al. 2002). Finally, because it is virtually impossible to collect enough data from older animals to illustrate the grandmother hypothesis adequately, data from human beings are considered, to see exactly how it works out.

SPERM WHALES

We do not know about the mothering abilities of whales in detail, as many of them range over thousands of kilometers annually, but data gathered from sperm whales indicate that their skills are impressive. Older females are important in "babysitting" the herd's young, most or all of whom are their relatives. Infants less than a year old have no trouble keeping up with the normal swimming speed of the herd (4 km/h or 2.5 mph), even when they are only a few hours old (Whitehead 2003). However, very young whales are unable to dive deeply, so when their mothers seek food near the ocean floor a kilometer or more below, they are forced to remain near or at the surface of the water. They hang out there for up to an hour at a time, with other infants and with protective females (some of whom are elders) other than their mothers. The mothers dive for food at different times, so their young remain well protected. (Whale young of most other species accompany their mothers at all times, but northern bottlenose whales, short-finned pilot whales, and other deep divers have patterns of "babysitting" similar to those of sperm whales.)

Older females not only protect young who are not their own, but may give suck to them as well. Observers have seen one calf suckle from two females at different times, and two calves of about the same size (and therefore not siblings) suckling from the same female. In one study of 22 older females between the ages of 42 and 61, none were pregnant or ovulating, but six were lactating (Whitehead 2003).

Whitehead also noted that older female sperm whales, along with other adults, help protect the young when the herd is threatened by predators such as orcas. If pod adults are spread out deep in the ocean, foraging for fish or squid, when orcas are detected, they stop clicking and silently swim swiftly to the surface. There they form a defensive ring, their bodies within a few meters of each other, the young in the center of their circle. If there is a male present, he may hover at the periphery, too immense, personally, to be vulnerable; he himself can injure an orca with a blow from his fluke.

The behavior of pod members varies when they are threatened. Sometimes all the individuals face outward, especially if there are many attackers. Sometimes they face inward, in what is called a Marguerite formation, because their tails fan out like the petals of a marguerite (a daisylike flower). If orcas dive below the circle, the sperm whales face downward, watching their enemies. None tries to flee or dive deeply to escape. Orca whales are as swift as they are and, with echolocation, they could determine where and when the sperm whale would surface in order to breathe, and wait there to attack. Although the protecting sperm whales are sometimes bitten or killed in encounters with orcas, the herd young are protected. It takes a sperm whale pod, or an elephant herd, to raise one of their youngsters, just as it takes a village to raise a child.

LANGUR MONKEYS

Langur grandmothers do not spend all their time fighting for the good of relatives in their troop; they also give support to their grandchildren. As Carola Borries (1988) pointed out, "grandmaternal behaviour provides the possibility for elder and even for post-reproductive langur fe males to further increase their inclusive fitness. Even the slightest reproductive advantage should be used by individuals and will be selected for [in evolution]."

Borries studied a troop of 18 langurs in Jodhpur, India, comprising 1 adult male, 13 adult females, and 4 young. Among the females were two rather scrawny grandmothers who were no longer reproducing. One of them (#6) had an adult daughter, a juvenile granddaughter, and an infant grandson; the other (#3), an adult daughter and an infant grandson. During 1019 contact hours with the troop, Borries noted various interactions between the grandmothers and the young (often at half-minute intervals), and how close they were together.

Borries found that #6 treated her grandchildren differently. She calmed the male infant to some extent during weaning episodes, and sometimes interfered aggressively when he was playing with other young or was near non-troop males, but she formed no social bond with him, as if she knew he would leave the troop when he grew up. However, #6 did form a social bond with her granddaughter, who sometimes approached her closely and with whom she had mutual grooming. This grandmother often interfered with activity between her granddaughter and other females (but not her mother), and in rough play with her peers. Her intervention perhaps informed other langurs that the granddaughter had a defender, and helped integrate her into the troop where she would spend the rest of her life.

Grandmother #3 surprised the researchers, because she preferred not her own granddaughter, but another female infant (possibly a close relative) who had some contact with her and received some grooming, at the youngster's instigation. This may be because her granddaughter was her daughter's overly-protected first baby, and the other infant was at least the seventh youngster of a less-protective mother. Grandmother #3 was remarkable in that she never showed special concern for any infant. Neither of these two grandmothers really mothered their grandchildren, perhaps because the infants' mothers were too protective of them (since they were of low rank in the troop), or because this activity did not interest them.

Borries considered the costs and benefits of the grandmother/grandchild relationship, the costs being the extra energy expended by the grandmother, and any advantage being to the benefit of the grandchild. The costs seemed small—mainly extra vigilance and occasional skirmishes by the grandmother (that could potentially harm her, but did not do so). However, the benefits to the young could be psychically great: a social bond to give comfort and encouragement, protection by a vigilant senior, and connection to a troop member other than the mother. Borries assumed that "grandmaternal behaviour really benefits the child," and concluded that by her activity, a grandmother is investing to some extent in the future of her grandchildren, even if this is not readily apparent.

HUMANS

Human grandmothers have always worried those scholars who ruminate on evolution. Not grandmothers themselves, of course, but the concept of grandmother. In most species, individuals do not live long once their reproductive life is over. In human beings, women stop reproducing after menopause, when they are about age 50. Hundreds of thousands of years ago, most humans of either gender did not live to be that old, but nowadays Western women usually live for decades past menopause, sometimes becoming not only grandmothers, but great-grandmothers or even great-great-grandmothers. Why?

One theory is that during human evolution, healthy older females helped their grandchildren by supplying them with extra food and care. These children were more likely to thrive than other children, and therefore pass along their grandmothers' healthy genes and wisdom to their own children. Over time, our ancestors began to live longer. Now they could afford to have children who took many years to mature, long years in which their brains developed and elders taught them how to cope effectively in their environment. The grandmother hypothesis of human evolution seems more probable than any other, including the hunting hypothesis, where male bonding by hunters—sharing meat, and venturing into new territories after prey—are paramount (Hawkes 2003; Dagg 2005).

Studies on "primitive" societies—such as the Tanzanian Hadza, among the last people to live as hunter-gatherers in Africa—show that elderly women in their fifties and sixties are wiry, fit, and incredibly active. They are useful in helping to care for grandchildren, but they also search for food that they bring back to camp, often far more than they can eat themselves, usually giving the extras to their relatives. They "forage the longest hours, dig deeper for tubers, and spend more time gathering berries and processing food than any other category of forager" (Hrdy 1999). When grandmothers supply food, it helps their grandchildren maintain their weight, grow faster, and be more likely to survive (Hawkes et al. 1997). However, older women in other tribes, such as the Aché and the Hiwi in South America, do not bring surplus calories to their groups (Peccei 2001).

Recently, statistical research has shown that as far as evolution is concerned, even grandmothers in Western society have remained important. The analyzed data came from two populations: (1) in Finland, from old Lutheran church records of people living in five farming and fishing communities, and (2) in Canada, from records of everyone (mostly Catholics) born in the region of Saguenay, Quebec, during the eighteenth and nineteenth centuries (Lahdenperä et al. 2004).

Both sets of women were hugely productive, with an average number of 6.8 children for the Finns, who in turn produced 11.3 grandchildren, and 9.1 for the Canadians, who produced 38.2 grandchildren. For each family in each group, the presence of grandmothers in their children's lives meant that there were two extra grandchildren for every decade the grandmothers lived beyond age 50. When granny was around to help out, her daughters had children earlier, her sons and daughters had more children who were more likely to live to adulthood, and these parents themselves lived longer and healthier lives.

Today, grandmothers (and grandfathers, too) remain important in many extended families: babysitting, reading to the children, cooking, helping with cleaning and laundry, and recounting apparently unlikely stories to youngsters about what life was like long ago. When I had babies, my mother was invaluable in telling me what to do about breastfeeding, diapers, burping, and crying (both the babies' and mine, from utter exhaustion), until I was able to cope on my own. I could not have managed without her.

Jared Diamond (2001) explicitly stated that human grandparents were so influential, they played not only an important, but indeed a decisive role in human evolution. He believed that after the development of language, older people increased the survival rate of their younger relatives by both helping them physically and sharing a lifetime's worth of experiences: what plants to eat in times of famine, what to do when cyclones struck, how to treat people felled by a pandemic, how best to evade (or incite) enemy raids. Such catastrophes, which occurred only at long intervals, could wipe out an entire tribe if it did not know how to react. The knowledge of older people then was the equivalent now to information in our libraries or on the Internet.

Diamond argued that the significance of the shared experiences of grandmothers was so great that it overrode the importance of the children who would have been born had the grandmothers not entered menopause. If they had continued reproducing, they would have been more likely to die from childbirth and lactation, traditionally dangerous activities whose danger increased with age, leaving the group in greater peril than it would have been otherwise. Diamond noted that "similar considerations may have driven the evolution of menopause in killer whales and pilot whales, which also live in lifetime groups with complex relationships and culturally transmitted survival techniques."

IN SUMMARY, FAR FROM BECOMING passive with age, many species include older females who are vital in improving the future of their offspring, and thus of their own genetic legacy. Their help may be of such importance in evolution that they live many years after they themselves have stopped reproducing. Old females do this either by fighting for their group, or by providing direct care for their relatives. Human statistics for grandmothers living before the use of contraceptives indicate that their presence and help may mean that the number of their grandchildren is increased. Senior animals who are still involved in reproduction, and lead sexy lives, are the subject of the next chapter.

CHAPTER 13

Sexy Seniors

REPRODUCTION IN BOTH HUMANS AND NONHUMANS decreases with age, as noted in earlier chapters. Females produce fewer or no young when they are older, and very old males may be impotent (such as wolves [Mech 1966] and red deer [Darling 1969]). But that does not mean that seniors necessarily lose their interest in sex. Some oldsters are incredibly lusty, such as "old goats," a label often and appropriately applied to licentious older men. Many human beings continue to practice sex into their seventies and eighties; if older people have partners and are in good health, their sexual activity is about as frequent as it is in other age groups (Lindau et al. 2007). Unlike human beings, however, animals generally only copulate when the female is in estrus or ovulating (bonobos are the exception). In large mammals, lactation may delay estrus in females for years, so observing mating in the wild is uncommon for many species. This chapter gives a voyeuristic look at sexual behavior, recorded by zoologists, of a number of older individuals of different species: chimpanzees, bonobos, gorillas, langurs, baboons, dolphins, lions, and meerkats.

Sex in the Wild

CHIMPANZEES

Flo, the female chimpanzee much admired by Jane Goodall (1986), was one of the lusty ones. In 1963, when she was at least 35 years old, she came into estrus following the five years she spent nurturing her daughter Fifi. She gave off such a powerfully sexy presence that she was followed by 14 males anxious to mate with her, and she did so with many of them. Sperm competition in chimpanzees means that females in estrus mate with many males, who each produce large amounts of sperm; the rivals' sperm then "compete" within the female's reproductive tract to determine which will fertilize the single egg (Birkhead and Møller 1998). The high-ranking male Hugo followed Flo everywhere during her five-week period of estrus, offering her comfort with a touch or a hug if she was hurt or frightened. He even remained close to her for a two-week period after her estrus ended, when all the other males had lost interest.

Four years later, Flo was again seductive. When she entered estrus after raising her son Figan to semi-independence, she mated 50 times in one day. Shortly before this period, Flo was at her peak, the highest ranking female in her troop, able to win fights with all other females (unless they recruited their sons to help them). If a female threatened Flo's young, she had to be prepared to fight Flo. Being in estrus made Flo even feistier than usual, and able to grab more bananas at the feeding station—which was fortunate, because her teeth were already worn down to the gums.

Inspired in part by Flo's passionate nature, three researchers from Boston decided to investigate whether older female chimpanzees in general were more sexy in male eyes than young prime females, preposterous as this possibility seemed, given the power of the media to valorize pretty young women in human culture (Muller et al. 2006). The data they analyzed, collected over a period of eight years (1996–2003), were for copulations observed in a chimpanzee community in Kibale National Park, Uganda.

To ensure that their findings were accurate, given the unlikeliness of their hypothesis, they examined the data in four different ways. Did males making copulatory advances (glancing with an erect penis, and branch-shaking) approach older females in estrus in preference to younger females? They did. Were males more likely to join groups that included females in estrus who were older rather than younger? Yes, they were. Did older females tend to mate with younger males, with high-ranking males, and with alpha males? Yes, they did. And in mating contexts, did males compete with each other more aggressively (targeted charging displays, chases, attacks) to have access to females in estrus in groups that included older females? Yes, they did. The researchers conclude (although rather reluctantly, it seems) that "chimpanzee males may not find the wrinkled skin, ragged ears, irregular bald patches, and elongated nipples of their aged females as alluring as human men find the full lips and smooth complexions of young women, but they are clearly not reacting negatively to such clues."

BONOBOS

Old Kame, like Flo, was extremely sexy. She was a bonobo or pygmy chimpanzee (*Pan paniscus*) who was, like all females who were not nursing a newborn, almost always in estrus, with a large genital swelling. (Common chimpanzees are only in estrus and willing to mate for short cyclic periods at a time.) Takayoshi Kano described Kame in his book *The Last Ape: Pygmy Chimpanzee Behavior and Ecology* (1992). She may have been elderly, but when she arrived at the sugar-cane feeding place set up by researchers in the Congo in Africa, she created a frenzy among both female and male compatriots. The males were keen to have sex with her, but so were the females.[1]

Female-female sexual contact is so common in bonobos that it is given initials—GG (for genito-genital) rubbing. If Kame was feeding, female X would approach her casually, linger a bit, and sit down beside her. Then X would make "a demand for invitation" by standing upright on two feet, extending her hand toward her friend, putting her face close to Kame's, and peering into her eyes, as if to say "Please associate with me in genital rubbing." Kame would roll onto her back, spread her thighs (the favored position used for ventral copulation with a male), and embrace X with her arms when X climbed onto her stomach (so that their genitals were near each other). Then, in this stomach-to-stomach position, they would rub their genitals together, clitorises touching. With their faces indicating "an uncontrollable emotion," they made quick, rhythmic thrusting movements, similar to those during copulation, but more side to side. The females ended this enterprise with a scream, similar to that emitted during copulation. (Although author Kano does not mention it, the word "orgasm" springs to mind.) Kano was convinced that the forward position of the genitals in female bonobos "evolved for GG rubbing rather than for ventral copulation."

Kame was not always interested in sex, though, because she and her family stayed away from the bonobo group and the feeding site after her 1-year-old abnormal daughter Kameko died. She allowed her young son Tawashi to carry the newly-dead body pressed against his chest, followed by her second son Mon, as she walked dispiritedly in front of them. The next day she stood sadly by the corpse, staring at it and shooing away flies. She was clearly depressed, and remained so for a long time, having few social relationships with others.

At puberty, bonobo daughters leave their natal troop to join a new group (de Waal 2005). Often an older female in this new troop becomes a mentor and protector for such an immigrant, with the two bonding by grooming and having sex with each other. Unlike young females, young males stay in their natal troop with their mothers, with whom they retain a strong bond throughout life. Originally it was believed that such a male/female duo was a couple, emulating human marriages, but this was not so. De Waal's research found that an adult male and his mother, who is somewhat smaller, travel and hang out together in the forest as if they are a pair, but they do not copulate. The son benefits from his mother's attention and protection, particularly if she is of high rank. "Instead of forming ever-changing coalitions among themselves, male bonobos vie for positions on their mothers' apron strings."

Kame had three grown sons, the oldest of whom was the alpha male of their group, in part because of Kame's high status. When Kame became frail as she grew older, she could no longer back up her children as strongly as she once had. The son of the beta female must have realized this, because he began to challenge Kame's sons. The beta female also began to confront Kame's oldest son, on behalf of her son. Discord escalated to the extent that the two mothers were hitting each other, rolling about on the ground with Kame held down by the beta female. Kame, humiliated at this defeat, never recovered her former status. Her sons soon dropped to mid-ranking positions in the troop. When Kame died, they became peripheral to the group, excluded by the alpha male, who gained this rank as the son of the new alpha female.

As is evident from this unusual relationship between mother and son, bonobo mothers gain an increasing sense of self-confidence as they grow older, as well as the respect of all the males. When an older mother and son approach the feeding station, the mid- and low-ranking males quickly get out of their way, while the alpha males usually turn their backs to protect their sugar cane morsels or flee on two legs, clutching food to their chests (Kano 1992). If Kame's sons had been chimpanzees, they would have banded together in defense of their positions. Such male alliances are poorly developed in bonobos, however, which allows older females to have significant power in the group.

Older bonobos in zoos are as sexy as those in the wild. Frans de Waal (2005) described Loretta, whom he had known for 20 years, recognizing him as he walked outside her enclosure, and greeting him with shrill hoots. She presented her ballooning genitals to him, gazing at him upside-down from between her legs, and waving at him invitingly.

GORILLAS

It may be that gorilla males, as well as chimpanzee males, are more smitten by older rather than by young females, although so far there are few data to prove this. By 1984, Effie, of Group 5, was no longer copulating solely with Beethoven, who was also getting on in years. She was now mating with two of his sons, Ziz and Pablo, without any complaint from Beethoven. None of these three large males were interested in copulating with "the sexually mature, hot and eager young females Puck, Tuck, Poppy, and Pansy," who were also members of group 5 (Mowat 1988). In a literature survey investigating the sexual preferences of 14 species of male primates (Anderson 1986), all these species had males who preferred to mate with older rather than young females (although the females were not necessarily very old). In general, these females were already successfully raising young, so the males' choices would be sensible, in that they would know they were inseminating females who were fertile and were competent mothers.

LANGUR MONKEYS

Older female langurs (unlike female chimpanzees and bonobos) can be highly aggressive, attacking males who invade their troop, while the prime females (along with their infants) step back and coolly observe the mayhem (see chapter 12, both for this and for the incident below involving Sol). However, when and if they come into estrus, their pugnacity may vanish. For example, Sol had a history of battling males, but when she was in estrus she joined in a consort relationship, keeping company for days at a time with her former archenemy Mug, who had tried to kidnap the infant that Sol was determined to protect. During this estrus period, she also solicited sex from and mated with at least three other males, for a total of seven times, although she apparently did not become pregnant (Hrdy 1977).

BABOONS

Older male baboons may be keen on sex, as was the baboon Sherlock who, at age 24 (or about 90 in human terms), was still a successful lover (Smuts 2000). Such oldsters can no longer win mating partners by their alpha status or prime physique, but they may be able to do so by becoming friendly with individual females. Barbara Smuts studied the same species (or race) of baboons as Robert Sapolsky Papio cynocephalus anubis—but at a different site in Kenya, near Gilgil. In her book Sex and Friendship in Baboons (1985), Smuts focused on the characteristics of male and female friendships in the Eburru Cliffs troop, where most females had one or a few male Friends (just as the males might have one or more female Friends) who spent part of the time sitting near them, and with whom they exchanged grooming. (The word "Friend" is capitalized here because it represents this special relationship.) This Friendship was not directly about sex, because usually the females were pregnant or nursing. Females tended to avoid most males—because males were twice as big and often aggressive, attacking them sometimes for no known reason—but with their male Friends they were relaxed and almost affectionate, even if their young were present. Sometimes a female left her infant with her Friend to go off to forage nearby by herself, knowing that it would be safe with him.

Male Friends tended to be older than prime males, because the former had been in the 150-member troop longer, and had had more time to interact with females and gain their friendship. When Smuts analyzed her observational notes, she found that on average it was older males who spent the most time in consort activity, which meant keeping company with a female when she was in estrus and mating with her. This description makes it seem as though the males have control of mating, but this is not the case. A male coalition may defeat a consorting male, but the female may bound off and join another male entirely. There are virtually no known cases of rape in social primates (Smuts 1992), perhaps in part because the females band together to keep the males in line.

The most successful Friend was old Cyclops, low-ranking among the males but the Friend of four females, and the most successful consorter of males of all ages. He was a calm individual. When high-ranking males threatened him at length, circling around him and flashing their canines within a few centimeters of his nose, he merely sat still, unmoved, until they tired of their inability to provoke a response and wandered away.

On 13 occasions during Smuts's research, coalitions of two to four older males took over consort activities with a female in estrus who had been with a prime male; one of these elders then joined up with the female to form a new consortship. Only once did an aging male take an estrus female from another male on his own, without the help of another elderly resident (Smuts and Watanabe 1990). Usually a lone older male would not have been able to defeat an alpha or prime male for the privilege of consorting. These elderly Friends usually did not lure a female away from other older males, perhaps because they sensed some sort of senior-male bond, or might want them as future allies. Friendship males were especially likely to mate with females at their peak period of ovulation, and so father their young. Older male baboons rarely extend a coalition into a close brotherhood, as Alexander and Boz did (see chapter 10); such a relationship is unknown among young prime males who see each other not as pals, but as rivals (Smuts 1985).

Smuts described a meeting one day between the aging Boz and Iolanthe, one of his six female Friends. Boz, along with a few other baboons, had separated from the main part of the troop. They were returning when Iolanthe, a young adult with her first infant on her back, noticed the small group approaching over the eastern horizon. She, and others, turned to stare at them. Long before Smuts was able to make out who they were with her binoculars, Iolanthe began to grunt excitedly, rousing the others to look up and begin to grunt, too. Iolanthe broke away from the rest to move forward a few meters, grunting even more intensely. Then, when the elderly Boz was about 100 m away, she rushed forward to greet him. When she reached him, he gently touched the infant on her back. Iolanthe turned around and walked side by side with Boz to rejoin the rest of the troop.

Elsewhere, Barbara Smuts (2000) described how Sherlock, a very old male, worked his magic. When he and a female sat near each other, if she suddenly glanced at him quickly twice in a row, Sherlock would start to concentrate on eating leaves. If she leaned slightly away from him, he took a step back, looking about nonchalantly. If she caught his eye and did not look away, he would grunt softly and flash a mild come-hither face, but not touch her. He would approach and touch her only if she eventually made a come-hither face back at him. In time, she would willingly mate with him. His recipe for success, which Smuts emulated in her work with all the baboons, was "carefully observe the other, express your friendly intentions, keep your overtures subtle, and wait for the fearful one to make the first move."

DOLPHINS

"Friendship" may pay off for dolphins, too. Over the years, some information on sexual behavior has been collected for older bottlenose dolphins in the wild; it is difficult, but not impossible, to identify aging individuals if one spends much time with them, in part because their white chins and genital patches enlarge with age (Herman 1980). Researchers noticed that older, lethargic males in Shark Bay, Western Australia, did not herd females as enthusiastically as did younger males, but they spent more time in friendly interactions with females (Connor and Peterson 1994). The authors theorized that perhaps older males were no longer able to compete physically with young ones, but by being "nice" to the females, they might attract at least one with whom to mate. Their advanced age indicated that they had a good genetic constitution, suitable for fatherhood. From what observers are able to deduce, many or most sea mammals have very active sex lives, and this propensity probably extends to older individuals. For example, one slaughtered, post-reproductive short-finned pilot whale had sperm in her reproductive tract, indicating that she had mated recently (Marsh and Kasuya 1991).

LIONS

Many lions in Africa are killed by poison, snares, guns, disease, accidents, or other lions, but a few, such as Leo and his pride-mates, survive to old age, males into their teens and females into their twenties. Older females are less likely to have cubs than their younger sisters, but some do reproduce. As mentioned in the introduction, M, an aging female in Nairobi National Park, was assumed to be sterile because she had had no cubs for two years, but then she bore a litter.

Older male lions are sexually active, too, although perhaps not up to the standard they had when in their prime—one young robust stud copulated 157 times with several lionesses over a period of 55 hours (Schaller 1972a). Terry Shortt (1975), who traveled extensively around the world collecting specimens for display at the Royal Ontario Museum in Toronto, recalled coming upon an ancient male near Lake Manyara in Tanzania. His blunted canines were yellowed with age, his thin, orange-red mane was straggly, and his hide was covered with innumerable scars, but with him were two young healthy females in their prime. The male was in a bad temper from the heat and flies, so that when Shortt's Land Rover approached him too closely, he snarled and charged the vehicle, giving it a swat with his paw. Then, with this successful retaliation, he stalked deliberately away into a bushy area, his tail lashing and the two females prancing after him.

MEERKATS

Sometimes an older male does not take the initiative for sex, but can be coaxed into a sexy mood by a female. Griff Ewer, one of my mentors while I was researching giraffe in Africa, noticed this in her study of meerkat (*Suricata suricatta*) behavior. A young female in heat, anxious to copulate with a 3-year-old male who was no longer very sexually active, would hover close to him, nipping at the tufts of hair on his cheeks. This usually led to a bout of semi-serious wrestling between them. "When the general level of excitation of the male had been sufficiently raised, he would attempt to mount" (quoted in Hinton and Dunn 1967).

Sex in Zoos

Sexuality in zoos can run the gamut. Some rare species are so upset by their captive life that they never mate (which is why they are rare in zoos). Others, especially monkeys and bonobos, relish the leisure time they have for frequent sexual behavior, making human parents likely to hurry their children past their cages. This section considers the sexual behavior of an older chimpanzee (Jimoh), an aging gorilla (Timmy), and a pair of old/young siamang gibbons.

CHIMPANZEES

Jimoh became the alpha male of the Yerkes Field Station chimpanzee group in Georgia, but he originally joined the group by chance (de Waal 1996). The staff had removed the previous alpha male from the troop and replaced him with two new adult males. The females, who had lived together for many years, attacked and badly injured them both, so that these males had to be removed. Several months later, the staff added two new males. One was soon rejected, but the other, Jimoh, was allowed to remain. Two older females had immediately approached and briefly groomed him, and one of these had fiercely defended him from an attack by the alpha female. Years later, records were discovered that showed that these three animals, the two females and Jimoh, had lived together in another institution long before. De Waal noted that "apparently this contact more than fourteen years earlier had made all the difference."

Jimoh was not a large animal, but he was an effective mediator at stopping quarrels among his colleagues, and was still enthusiastic about sex. Even though he was top dog, so to speak, one day a female favorite of his secretly went off to mate with the adolescent male Socko (de Waal 1996, 2005). Normally, Jimoh would have merely chased off the interloper, but since this female had repeatedly refused to have sex with him that day, Jimoh was enraged when he found them together. He rushed at Socko at top speed, chasing him all around the enclosure while Socko fled in terror, screaming and defecating. Before Jimoh could catch his adversary, though, nearby females began to bark—"woaow, woaow"—in protest. The callers looked around to see if other females would join the protest, and when they did—everyone, including the alpha female, all staring at Jimoh—their calls turned into a deafening roar. Jimoh stopped the chase in response to this cacophony, gave a nervous or even embarrassed grin, and the females went back to their usual activities. Jimoh was important in the group despite his age, but not omnipotent.

GORILLAS

Zoos can be heartless in their desire to have their charges reproduce, moving individuals from one zoo to another without regard for their possible feelings. After 30 years of solitary confinement in a cage in the Cleveland Zoo, a lowland silverback gorilla called Timmy was finally given a partner, Katie, whom he liked (Kaplan and Rogers 2000; Couturier 2005). Other females had been introduced to Timmy's cage over the years, but he disdained them, so they were soon removed. Katie was different. As soon as she and Timmy were introduced, they got along famously. They became devoted partners, spending much time together and going to sleep in each other's arms. Miraculously, it seemed that Timmy would at last find happiness at the end of his lonely life.

Alas, it was not to be. Katie did not conceive, so the zoo authorities, in their wisdom, decided to move Timmy to the Bronx Zoo, where he would have several females as possible mates, and to give Katie a new male companion. This decision brought huge complaints from the public, which were aired on the radio and in newspapers. How could the zoo move Timmy when he and Katie were obviously in love? Why did zoos feel that their agenda overrode that of Timmy, who obviously wanted to remain with his Katie? (A letter purportedly from Timmy himself was published in the newspaper.) Why did a decision by zoo officials prevail over the wishes of thousands of Cleveland's human citizens?

The case finally landed in court. Timmy lost. He spent his last days sitting on a rock in the Bronx, staring blankly into space. Back in Cleveland, Katie fought with her new partner, who attacked and injured her. The cruelty in separating Timmy and Katie, when they were obviously content with each other, continued to enrage the public. The zoo director excused his actions, stating that "it sickens me when people start to put human emotions in animals. . . . When people start saying animals have emotions, they cross the bridge of reality." Surely he is in the wrong profession?

GIBBONS

Animals can be sexy not only directly, but also by behaving in secondary sexual ways. For example, siamang gibbons (*Hylobates syndactylus*) are usually monogamous, unlike the other larger apes. A male and female form a close family bond with their offspring—one infant is born each year, but each youngster needs three or four years of parental guidance before leaving family life.

Observers at the Louisiana Purchase Gardens and Zoo were amazed when their 18-year-old male was immediately sexually excited by a 6-year-old female, newly added to the zoo (Maples et al. 1989). He had been captured in the wild as an infant, lived as a couple for five years before his partner died, and then spent the next twelve years alone.

Siamang pairs in the wild announce their partnership by singing loud and elaborate duets. Susan McCarthy (2004) described such a duet.

> [The male gives two deep booms, the female then booms once, the male booms twice again,] and the female must immediately come in with accelerating high-frequency barks. After about the fifth bark, the male should utter an ascending boom, the female's barks should speed up, and the male should do a bitonal scream. At once the female must start another series of barks, and after five the male should do a scream, this time a ululating scream. The female does some fast high barks, then both of them bark and hurtle about.

The Louisiana Zoo pair began to practice this duet right away, and their performance improved quickly. Over a three-month period, the percentage of started great-call sequences which were correctly completed rose from 24 to 79 percent, and finally almost all were perfect. "One of the most common mistakes was for the female to start her first set of accelerating high barks without waiting for the male's second double boom. The other common mistake was for the male to bungle his bitonal scream, giving either the ululating scream or the locomotion call instead."

Why do pairs of animals—including many species of birds (Carrighar 1965)—sing duets (Alcock 1988)? Perhaps it enables each to assess the behavior and vigor of the other as a suitable mate. Maybe it serves to alert each to where the other is in dense vegetation, and helps to prevent mating with other individuals. Possibly it informs other animals in the area that a particular territory is taken by two individuals, who will defend it. Sex and reproduction are related to all these possibilities.

IN SUMMARY, AS FAR AS SEXUAL BEHAVIOR is concerned, older (but not extremely old) animals are often no different than their younger comrades. Both have emotions, just as human beings do, despite the disparaging comment of the Cleveland Zoo director. Some older males will mate with any female they come across, given her acquiescence. The aging gorilla Beethoven chose to mate with some older females in estrus, but not with "hot and eager" young ones. Some females in estrus are choosy about whom they will partner with, while others, such as the chimpanzee Flo, will mate with as many males as are willing. It seems that variety is the spice of life for oldsters as well as the young.

CHAPTER 14

Their Own Person

OLDER ANIMALS WANDER ABOUT ALONE for various reasons: they are loners by nature because of their species; they become eccentric over the years, like the unneighborly moose described below; they have slowed down with age and can no longer keep up with their group; or they have been expelled from their group—as were female langur monkeys and Nelson the hyena—because they *are* elderly. Each older animal noted below has been admired by a human observer who felt that it was important to preserve his or her individual character.

Mozu

Mozu is an amazing example of adaptability and persistence in a Japanese monkey. She lived in Jigokudani Park in the Japanese Alps, an older female born with no hands and no feet, congenital malformations attributed to pesticide poisoning (de Waal 1996). She moved by propelling herself along the ground with her stumpy limbs, trying to keep up with her colleagues. The others leapt nimbly from tree to tree in ice and snow, while she forced her way through high drifts, often with an infant on her back. Yet over her lifetime she produced and raised five young, none of whom had physical abnormalities. Fortunately, the park had hot-water springs where Mozu could warm up, and feeding stations where she could obtain sufficient food beyond what she found in the wild.

In 1991, when she was 18, her troop of over 200 animals was so large that it split in two along maternal lines. The more powerful division, which included the dominant females and their families, began to claim the feeding sites for themselves. The other division of subordinate females, to which Mozu belonged, was therefore marginalized. Mozu had to decide what to do. If she were prevented from visiting the feeding sites, she might not be able to find sufficient food to survive, but if she left the subordinate group, she would be leaving all her relatives, including her own young.

Mozu opted for sure survival. She began making overtures to the females in the dominant group (with whom she had grown up) by "grooming" them—gently pawing them, because of her stumps—as well as she could. Some of the females occasionally attacked her, but Mozu remained resolutely at the edge of the group, making herself inconspicuous. Eventually some of the females began to groom her in return. Finally, she became so well integrated into the dominant troop that she was allowed to forage at the provisioned sites. Life was good. I wonder if, while she was enjoying the bounty offered by human beings, she ever thought back to her birth family and her own young, from whom she had had to separate herself in order to survive.

Hugo

Hugo the chimpanzee, like all older males of this species, both began and ended his life as a low-ranking individual (Goodall 1986). Although large, he was never an alpha male. Yet by his behavior, he was sometimes a leader, guiding his troop from one place to another. When he first saw Goodall he threw a rock at her, rather than fleeing as most chimps did. He missed. He was bold but not overly aggressive—except around bananas, which he loved. Once, when he had caught a large infant baboon, he sat nibbling its flesh by himself for nearly nine hours while his friends waited optimistically around him, hoping for a tidbit. His teeth were worn and his body shrunken as he grew older, and he was often alone, ignoring social gatherings of the others. However, he and five other males did take part in a lethal attack on Godi, a male from a break-away group of Hugo's troop, with Hugo biting him and throwing a 5 kg (11 lb) rock at him (which missed). He and his aging grooming mate, Mike, died in the same month during a pneumonia epidemic.

Nelson

Loners are often social animals who either lagged behind or were spurned by members of their group. As they live a quiet life, all on their own, I like to think of them as philosophers, ruminating on their past or future life, and enjoying the peaceful present. Nelson the hyena, in the Ngorongoro Crater of Tanzania, was one such loner. He had a hard time of it, not only because he was older, but also because he was a male and thus subordinate even to low-ranking females. (Hyenas, unlike most mammals, live in a fiercely matriarchal society.) If a kill was small, the females sometimes prevented him from sharing it, so he then went out alone on foraging trips (Lawick-Goodall and Lawick-Goodall 1970).

Nelson's ears were tattered from past spats over food and females, and he was blind in his right eye. He had to walk with his head twisted to the right so he could see where he was going. Even so, he often tripped over tufts of grass or bumps in the earth. Once, when he was near a den, he stumbled and fell headlong into its entrance; had mothers been nearby, they would instantly have attacked him to ensure he would not harm their cubs.

Nelson, like all hyenas, often regurgitated hairballs. He loved to roll in the vomit, smelly as it was, and pick out undigested bits of bone; such fragments are already partly dissolved, which makes them a hyena delicacy. Nearby cubs liked to roll on and chew his hairballs, too, and they sometimes swarmed him when they saw him heaving and coughing them up, grabbing a prize hairball before he could prevent this. It made Nelson seem a bit of a loser, running away to escape from pursuing cubs so he could regurgitate in peace.

Jane Goodall was enchanted by Nelson, because he had an unusually pure baritone voice that she could recognize as his, even if she could not see him. Hyenas are notoriously noisy, with the members of a clan recognizing each other by their distinctive calls, and Nelson was especially vocal. He let loose the species' weird "ooooo-whup" whoop on every possible social occasion, so that everyone else among the clan knew where he was. Sometimes he walked along ooooo-whuppping in a quiet voice; perhaps he enjoyed listening to himself.

Raccoons

Older raccoons may also be musical. Sterling North, in his book *Raccoons Are the Brightest People* (1966), described a male raccoon who visited one of his neighbor's houses in New Jersey; this oldster liked to listen to their music, especially that of Bach and Beethoven. While North and his wife were visiting these people one evening, the host put on a record of Beethoven's *Ninth Symphony*. Soon after the music began to drift into the nearby woods, this elderly raccoon softly opened the screen door and padded over to sit facing the hi-fi speaker. When the last movement ended, he slipped out the door to return quietly to the forest. Joanna Burger (2001) had an older parrot who loved music, too, singing along and mimicking it whenever he could; Elma Williams (1963) mentioned domestic animals—goats, horses, and hens—who approached to listen to music played on a radio, and then dispersed when the music was replaced by an announcer's voice; and Bekoff (2007) cites a 45-year-old captive elephant who enjoyed Mozart's music.

Raccoons in the wild usually live for 10 or 12 years at most, by which time their teeth have become completely worn down (North 1966). However, if they are given a special diet, they can last far longer. Jerry, the pet of an elderly married couple, lived in a large cage for 22 years. Each day the wife baked a pan of cornbread, cut it into small squares, and fed him these, along with warm milk and a few drops of cod-liver oil. He died quietly and without pain, "perhaps the oldest, best-fed raccoon that ever lived."

Patches

Patches, a 12-year-old springer spaniel, preferred solitude to living among people (Redekop 2006). He lived east of Winnipeg, beside the rural house that his human companions had abandoned six years previously. Good-hearted neighbors scooped him up, planning to adopt him, but as they drove away, "Patches went crazy, tearing apart the van," so they quickly brought him back. Other friends built him two doghouses, but he refused to enter either of them; in winter he hunkered down under the house or in an old shack. People visited and brought him food and water, but he needed and wanted no permanent company. His days were spent quietly sleeping, sitting on a pile of logs to watch cars go by, or hunting for rabbits in the woods.

Wolves

Wolves often become loners because of an injury that prevents them from living in their normal fashion, within a pack. A few were so successful at killing sheep or calves for sustenance that they were infamous by the time they grew old. Lefty, of Burns Hole in Colorado, became a lone rogue after his left leg was caught in a trap (Bass 1993). He succeeded in breaking free only by twisting his foot off. His stump finally healed, but it left him with a distinctive gait and what some have considered a passion for getting even with the humans who had caused him such harm. Over eight years, Old Lefty was blamed for killing 384 head of livestock.

Almost as famous as Lefty were the Syca Wolf of southern Oregon, Three Toes of Harding County in South Dakota, and the Ghost Wolf of Montana. These western wolves had all turned to killing livestock—easy prey—because they had been injured by traps or been shot, and could no longer catch wild animals. As they aged, they were incredibly crafty in their pursuit of meat, for years evading ranchers, hunters, trappers, poison, helicopters, and airplanes.

Wolves are considered elderly at about age 10 (Mech 1966), although they can live to be 14 in the wild or 16 in zoos. By then they usually have worn or broken teeth and often travel alone, scavenging meat scraps where they can.

Old Mose

Grizzly bears (*Ursus arctos*), like rogue wolves, have also had a negative press; until recently, most older individuals were noted only for their infamy. Old Mose was named for his habit of moseying along at his ease (Petersen 1995). Harold McCracken described him in his book *The Beast That Walks Like Man: The Story of the Grizzly Bear* (2003). The "story" in the subtitle refers to the many ways in which and places where grizzlies have been slaughtered over the years. Old Mose had no compunction about tearing down settlers' fences when they restricted his movements. During his long life in Colorado, he was said to have killed about 800 head of cattle and other livestock worth over $30,000. He also killed one (or perhaps five, since exaggeration seems to be rife about Mose) of the many men who stalked him to earn the large reward offered for his death. These bounty hunters searched a broad area—120 km (75 mi) in diameter, straddling the Continental Divide—where the bear roamed.

Enos Mills told McCracken that Old Mose had a sense of fun. Several reports said he silently approached a camp where prospectors or campers were resting for the night, and then rushed through the campsite with mighty roars. Some claimed that he enjoyed the sight of men scattering for their lives and trying to scramble up trees like monkeys; perhaps he just wanted human beings out of his domain. He was said never to have harmed any men who were not trying to harm him.

In April 1904, when Old Mose was about 35, he was finally cornered by a large pack of well-trained hunting dogs soon after he had emerged, hungry, from hibernation, and was shot four times (Petersen 1995). His teeth were sound, his pelt was in good condition, he weighed over 450 kg (1000 lb), and the consensus of those present at his death was that but for their great victory, he would have gone on killing stock for several more years.

Beelzebub

Predators are not the only older rogue animals. Terry Shortt, in his book *Not as the Crow Flies* (1975), wrote about an unneighborly moose who hung around outside a cabin that he and a fellow researcher had rented near a marsh in northern Ontario. The elderly, swaybacked animal had shreds of skin hanging from his "less-than-imposing" antlers (it was early fall), a gaunt face, and a mangy pelt. The moose first attacked their clothesline, tearing off underwear, trousers, and socks and trampling the saliva-sodden articles into the tall grass. When an annoyed Shortt saw this, he advanced toward the animal, shouting expletives and expecting him to run away. Instead, the moose turned slowly around and marched purposefully toward him. Shortt, somewhat shaken, retreated quickly into the cabin and slammed the door behind him.

After that, when either man left or returned to the cabin, Beelzebub (as they named him) would paw the ground, snort, roll his eyes, shake his antlers, and chase them away from the yard he considered his own. The men learned to hurry through either the front or the back door, whichever was opposite to where the moose was at the time. Once, sick of always having to hide and then sneak into his own house, Shortt threw stones at Beelzebub, hitting him on his body and antlers. The irritated moose retaliated by speeding up his chase.

When their job of collecting specimens for the Royal Ontario Museum was completed, the driver of the truck they hired to take their belongings to the railway station refused to get out of the cab, because Beelzebub hovered nearby. He backed the vehicle right up to the cabin's front door, so they could load their boxes directly into the back of the truck. Then they, too, climbed into the back. When they were about fifty meters down the road, the driver stopped the vehicle so the men could join him in the cab without interference from Beelzebub.

Clarence

Clarence was an abandoned sparrow (presumably a house sparrow, *Passer domesticus*) who was unable to fly and so was hand-raised by a woman, Clare Kipps (1953), who taught him a number of tricks. He had an amazing memory. When he reached the venerable (for a sparrow) age of 12, he had a stroke. Kipps realized that she had no record of the act he last performed six years earlier, so she gave him his cues, and he remembered exactly what he had been taught so long ago.

During the London blitz in 1941, when he was young, Clarence's performance had enthralled children and others who were stressed out by the bombing. He would entertain them by playing tug-of-war with Kipps, selecting cards from a deck, and twirling a card in his beak. As his grand finale, "Kipps cupped her hands, and when she called 'Siren's gone!'—the signal to take refuge in a bomb shelter—he hid in her hands. After a moment he peeked out, 'as if enquiring if the All-Clear had yet sounded,' which always brought down the house." When the war ended, Clarence gave up show business (McCarthy 2004).

Dan

It seems fitting to end this chapter with a story that my father, Harold Adams Innis, an eminent Canadian economic historian, wrote in his autobiography, about his father. William Innis, born in 1865, was a farmer in rural Ontario.

> [Old Dan, his first horse,] had been ill for a considerable period and one morning when led out of the stable for a brief period he beckoned to my father to lead him through the gate leading to the back of the farm. My father followed and in turn the horse continued to indicate

his wishes that other gates should be opened and finally arrived at the site of the old homestead two miles to the south. The two of them then returned, the horse seemingly satisfied with the trip and after coming back to the stable died within a few hours. This was a tribute as much to the sympathetic interest of my father as to the intelligence of the horse.

OLD ANIMALS DO NOT NECESSARILY hang around, waiting to die. These anecdotes indicate that many remain as their own "person," living a satisfying life, no longer frisky with youth but alert and individualistic in their seniority.

CHAPTER 15

Adapting and Not Adapting

S OME OLD ANIMALS READILY ADAPT to novel situations, but other individuals and groups may find this impossible. Presumably this difference relates in part to how things are processed in the animals' brains. Even though turtles and tortoises have existed for a very long time, a particular member could come up with unique responses to unusual events. Individual primates (gorillas, chimpanzees, and bonobos) can be highly flexible in their behavior, as can elephants and dolphins. However, under artificial conditions imposed by humans, other individuals (for example, among dolphins and chimpanzees) may be unable to adapt completely, and whole pods of whales (pseudorcas) have been destroyed because of this inability to adjust.

Adapting

TURTLES

Metaphorically, you can teach an old dog new tricks. Older animals, even members of an order of reptiles that has existed for nearly 200 million years, may absorb new experiences. Pigface, a Nile soft-shelled turtle (*Trionyx triunguis*), spent almost his entire life at the National Zoo in Washington DC (McCarthy 2004). For his first 40 years, Pigface was kept in a barren enclosure, with his only stimulation being the pursuit of the 12 live goldfish he was fed once a week; he chased them "with speed and agility" until he caught and consumed them all. This took twenty minutes.

Pigface later injured himself by biting at his front legs and raking his neck with his front claws. Worried by this behavior, his keeper decided to give him some toys. This sensible act turned his life around. He began to nose and bite a basketball and push it around the tank. A floating hoop was even better: he could "chew it, shake it, pull it, kick it, and swim back and forth through it." He began playing tug-of-war with the hose the keepers used to refill his tank, and he loved to have the stream of water cascade over his head. Pigface had learned to play! He died a few years later, in 1993, when he was over 50 years old.

Another relative, this time a land-dwelling older tortoise at a menagerie in Mombassa, Kenya, also adapted easily to a new experience—the adoption of a 300 kg baby hippopotamus who had been orphaned when swept out to sea by tsunami waves in the Indian Ocean in December 2004 (Hatkoff et al. 2005). Although the tortoise, Mzee, was about a century old, 100 times as old as the newcomer, he quickly befriended Owen, the infant, who normally would have stayed with his mother for two or more years. The two animals swim, wander about, and eat and sleep together, their bodies touching. Despite his advanced age, the tortoise was quick to adapt to his new role as a guardian.

GORILLAS

Numerous captive silverback gorillas have become locally famous by the time they grow old, but none is more famous than Jambo. Not for what he did, but for what he did not do to a little boy. He is so well known that Richard Johnstone-Scott (1995) wrote a book about him.

On a hot August day in 1986, Levan Merritt, a young lad anxious to see the family of gorillas at the Jersey Zoo in Great Britain as closely as possible, leaned too far over the rail above the gorilla cage and fell over 3.5 m (12 ft) into the dry moat below. The crowd gasped in horror. Everyone watched transfixed as the aged Jambo, followed by his family, ambled over to the boy to see what was up. This looked like a young human. Why was he lying so still? Jambo leaned over to sniff the body carefully, shifting so that his massive form stayed between the boy and the other curious gorillas. Then he sat back on his haunches for nearly a minute, examining the throng of horrified faces that were staring down at him and yelling "Jesus Christ!" "Oh my God!" "Get away from him!"

Jambo turned again to the boy, gently lifting part of his clothing and brushing a dark finger along the pale exposed skin. Then he lifted his hand to his nose to smell the scent. Soon the boy began to gain consciousness, whimpering and struggling to move. At this, the murmur from the human chorus above became a babble of shrieks: "Don't move!" "Keep quiet!" "Watch out!" "Do something!" "Where are the keepers?" Startled by the chaos above him, Jambo turned away and, followed by the others, retreated anxiously to the gorilla house, where the doors were quickly locked.

Keepers rescued the boy, while the watchers recounted what they had seen again and again to others who had seen it, too. Jambo was no longer a threat, but a hero! He had shielded the helpless lad from the other animals, and led them away, lest they hurt him! Levan recovered completely and was himself celebrated, as the boy who was saved by the wise old gorilla.

Jambo became an icon for his species, appearing in various television programs. The Fauna and Flora Preservation Society used his image to remind people that gorillas were not the bloodthirsty creatures conjured up by movies such as King Kong and by hunters wanting to justify killing them; gorillas were peaceful vegetarians best left to live out their lives in the jungles of Africa.

Jambo was famous even before Levan's fall, because he was the first gorilla born in captivity to be raised by his mother rather than bottle-fed. He was a model silverback, respectful of, and respected by, the other gorillas. He could be pugnacious, but also incredibly gentle. Once he caught a starling who was stealing his food; after examining it carefully, he let it go. Another time he flung up a woman's eyeglasses that had fallen into his enclosure, so the keeper could catch them and return them to their owner. Jambo liked to play with water from a hose, washing his hands and feet while drenching everything around him. He loved to be tickled, his eyes shut tight, his jaws wide open, giggling wildly.

When he died of an aneurysm in 1992, shortly after the birth of his fifteenth grand-offspring, Jambo weighed 180 kg (400 lb) and was 1.78 m (70 in) tall when he stood upright. Fittingly, he was memorialized by the Jambo Award, given to individuals or to a group that had made humans "more aware of the true nature and/or needs of the other animal species with which we share the world." The first £100 award went to the Digit Fund (now the Dian Fossey Gorilla Fund), established by Dian Fossey to honor Digit, the mountain gorilla who had been butchered by poachers.

Willy B., a 41-year-old male gorilla at Zoo Atlanta, devised a game with his keeper, Charles Horton (Linden 1999). If Willy noticed that a visitor was chewing gum, he would come to the front of the enclosure and chew. When the visitor asked what Willy was eating, Horton would reply that Willy had nothing in his mouth, but wanted a piece of gum to chew. Most visitors agreed to give him a piece. Willy would then walk over to where food was delivered to the gorillas, and wait for Horton to come there, bringing him the gum.

CHIMPANZEES

Heini Hediger (1968), a past director of the Zurich Zoological Gardens, knew that apes and monkeys sometimes held drinking water in their mouths to squirt at people who came near them. He was cautious when, after taking pictures nearby, he had to walk past the cage of an old chimpanzee. Hediger noticed that the animal was sitting listlessly with his back to the walkway, apparently playing with his toes and taking no notice of him. He looked at the chimp's cheeks, but decided that they did not bulge with water. However, when he was opposite the cage, the chimp swung around in a flash, leapt to the front of the cage, and drenched Hediger with warm water from nearly 3.5 m (12 ft) away. The trickster must have collected water just in case Hediger should pass his cage. As an animal psychologist, Hediger was embarrassed to be so caught out, but pleased that the ape's joke was successful. His dousing also underlined the chimp's intelligence in thinking up such a prank, as well as the need that captive animals in general have for toys they can use to amuse themselves.

An elder, male bonobo used his head to save the lives of his cage-mates. One day at the San Diego Zoo, Kakowet suddenly came to the keepers' window, screaming and waving his arms to catch their attention (de Waal 2005). The moat around the bonobo enclosure, 2 m (6.5 ft) deep, had been drained for cleaning and the keepers were about to refill it with water. But Kakowet, who knew the cleaning routine, had noticed young bonobos trapped in the moat and unable to get out. A keeper heeded Kakowet's warning, saw the juveniles, and provided a ladder so they could climb to safety. Kakowet himself pulled out the smallest youngster. All were saved.

ELEPHANTS

Both African and Indian elephants have trouble when they come into contact with human activity in the wild, because they have been able to adapt only too well to farmers' food sources. As human populations have expanded in India, many of the elephants' home forests have been cut down and replaced with agricultural crops. Older adult animals remember enjoying plants that flourished near villages—bananas, mangoes, coconuts, and sugar cane, as well as grain crops. Elephant raids on such crops, especially by males, can be frequent and devastating. One older male arrived at a cultivated field at 6 p.m. and gorged on millet steadily for twelve hours, consuming about 250 kg (550 lb) of grain. Fences are not strong enough to stop them, and even men keeping guard beside fires may be ineffectual, because the Hindu religion forbids them from shooting elephants (Chadwick 1992). About 150 to 200 people are killed each year by wild elephants.

DOLPHINS

Bottlenose dolphins are wary of human beings, with good reason. Kenneth Norris (1974), in his desire to study dolphins in captivity, decided to have his ship, the *Geronimo*, approach a school of them off San Diego Bay, so he could capture some with nets. One of these dolphins was Scarback, readily identifiable to human watchers because he had a dorsal fin and tail with regular deep cuts, healed over and now white, from a much earlier collision with a ship. He had come to know the *Geronimo*, too, having previously jumped over a net (something few dolphins did) that Norris had set out, and rammed right through another.[1]

Day after day, Norris set out to stalk Scarback's school, but this elder sage quickly realized what was going on. Whenever the *Geronimo* came close, he rose to the surface of the water, looked at the ship with one eye, and then the school departed. He knew the *Geronimo* was an enemy, either from its appearance or, more likely, by the distinctive sound of its engine. "It's no use," the ship's captain told Norris. "We'll never get near them. They'll be riding the bow waves of that tug over there before you know it." Sure enough, and much to their annoyance, there were Scarback and his pals, surfing beside the tug without needing to move a muscle. Half of the 100 bottlenose dolphins living in the warm Florida waters in and around Sarasota Bay are fairly old individuals, identified by radio transmitters or photographs from 30 years earlier (Wells 2003). A few females live into their fifties, but most males, who are considerably bigger, die before they are 40. The females continue to reproduce their entire adult lives, with some giving birth at age 48 and rearing these calves for several more years. The older a mother, the more experienced and successful she is in raising her young; the firstborn calf of a young female seldom survives to adulthood.

Outsmarted by Scarback and his group, Norris next tried to capture bottlenose dolphins in the upper Gulf of California in Mexico, where there were many large schools with "fat old patriarchs, ten to twelve feet [about 3.4 m] long and weighing eight hundred pounds [365 kg] or more." In these waters, the dolphins frolicked among shrimp trawlers, waiting for a free meal when the shrimpers threw unwanted fish overboard. But here, too, the dolphins were too clever to be netted, perhaps led by those "fat old patriarchs."

Despite Norris's difficulties, many dolphins have been captured for research. Investigators formerly took a dolphin from the wild and kept it captive for life. But what harm had the animal done to be so imprisoned? Carol Howard, in her book *Dolphin Chronicles* (1995), described a more recent trend of sensitivity toward these intelligent animals. After dolphins have been captured and studied for several years, they are returned to the ocean. These individuals are then carefully monitored for a few years, and seem to be able to take up their lives where they left off. One hopes they will live normally for the rest of their long lives.

Unable to Adapt

Some animals are so traumatized by captivity that they are unable to cope with what is asked or expected of them, such as Alice the dolphin and Vieille the chimpanzee. If a group of mammals or birds has never been exposed to a situation, then the older, more experienced leaders—and their followers—may or may not be able to respond suitably. It depends in part on what is hard-wired in their brains. If, during its evolution, a species has *never* been exposed to a situation and therefore has no ingrained response, there may be problems. Human activities, which now encompass the globe, have caused many of them. For example, when pioneers in the American West began to put up fences, changing the grasslands into agricultural fields, groups of pronghorn antelope found their centuries-old ranges constricted. These animals risk snapping the long, slender bones in their lower limbs if they try to jump, and many died if they were unable to find a way around the fences (Byers 2003).

Some animal predicaments, such as those faced by pseudorcas (also called false killer whales), are catastrophic rather than just troublesome. Pseudorcas, like all dolphins and whales, have large brains, but this does not mean that they are necessarily super-intelligent. In part they need a large brain to house systems dealing with their aquatic environment, especially those supervising sophisticated locomotion and sending, receiving, and interpreting echolocation sounds.

ALICE THE DOLPHIN

The captive older bottlenose dolphin Alice was skittish and shy; she had lived in freedom for much of her early life and was frightened by change (Norris 1974). The dolphins in one pool routinely had to be herded into a connecting one, so the first pool could be cleaned. No matter how often it happened, this was a traumatic experience for Alice. A net had to be drawn up behind her to push her in the right direction. As she neared the gate, which was 2.4 m (7.9 ft) wide, she lay on her side, calling out in despair. When she was about to be forced through the opening she would lunge back in hysteria, sometimes becoming entangled underwater in the net. Fearing she would drown, researcher Kenneth Norris had to leap into the pool beside her and haul her 180 kg (400 lb) body to the surface so she could breathe. Then, while she whistled in terror, he pushed her into her new quarters. She was so upset by this performance that it was several days before she could again carry out experiments. Far from being subnormally intelligent, Alice was a good candidate for important research on how dolphins "see" to catch fish at night or in murky waters. As a middle-aged adult, the American Navy captured her for research purposes. When she was older, Alice was loaned to zoologist Kenneth Norris for experimentation. Because of her age, which was well into her thirties, her **dorsal fin** was frayed at the edges, her skin streaked with scars, and her **body lumpy**.

In the early 1950s, only one "oceanarium" existed in the world, in St. Augustine, Florida. Few people were interested in whales and dolphins. No one knew if they would live long in captivity, or that they could be trained to do tricks. No one realized that dolphins use echolocation to catch fish in total darkness, or that the many different species of dolphins and whales have diverse social organizations (Norris 1974). Aside from whalers' reports, almost nothing was known then about the behavior of whales and dolphins in the wild, and for decades whalers continued their annual slaughters of tens of thousands of these animals.

Norris had been hired to help plan and build a second oceanarium, called the Marineland of the Pacific, in Los Angeles. He was anxious to exhibit marine mammals in the oceanarium, and later decided to do research on them when he became a professor at the University of California, Los Angeles.

The research project that Norris and his colleague Ronald Turner devised asked "How fine a discrimination can a porpoise (a synonym for dolphin) make using sound alone?" Despite her advanced age, Alice was well able to answer this question. After she was "blindfolded" with rubber suction cups fitted over her eyes, she received a reward of fish if she used her snout to press a lever in front of the larger of two sphere hung in the water. If she made a guess, she would be right half the time, so they repeated the test many times, with the balls randomly changing positions to make sure she was choosing size and not the left or right side. Over the course of the lengthy experiment, they gradually hung spheres that were more nearly alike in dimension, to test her ability to distinguish between them. For the experiment itself, the researchers chose a series of ball-bearings, from between 6.35 cm (2.5 in) down to 1.27 cm (0.5 in) in diameter.

At first Alice did not want to wear the blindfolds. To persuade her to do so, Norris and Turner declared a "time out" every time she balked, walking away from the pool and out of sight. Alice became upset at their desertion, shifting herself bolt upright in the tank and peering over the pool rim to see where the men had gone. After 20 more seconds of this "punishment," the men returned and Alice was ready to cooperate.

By the end of a year, Alice was more or less into a routine, and performed for many days without a hitch. After receiving her blindfolds each morning, at an underwater signal she would swim in a large circle, rise to the surface for a breath of air, submerge, and then swim toward the apparatus while emitting a train of clicks which were picked up by a hydrophone. If she pressed the lever in front of the smaller ball, there was no sound signal from the automatic feeder, so she jerked away in annoyance and began to circle again. If she chose correctly, the feeder clicked and a fish fell into the water, which she immediately wolfed down. Then the balls would be lifted out, their positions changed (or not) according to a sheet of random numbers, and re-lowered for another test.

When Alice had perfected her choice between two balls, she would be presented with balls that were closer in size, and the tests would resume. She made many errors at first, but then became more accurate in her choice. When same-sized balls were hung, which would return only one pattern of echoes to her, Alice would not push either lever, as if to say that it was impossible to choose between them. Eventually Alice showed that 77 per cent of the time—well above random choice—she could differentiate between balls with only a 1.91 cm (0.75 in) difference in diameter. The experiment was ended at this point, as the Navy wanted their dolphin returned. Alice had opened the eyes and minds of scientists, and demonstrated for the first time that dolphins were brilliantly adapted to an aquatic existence.

Eventually Norris decided to buy Alice, because she had become dear to him despite her foibles. He flew her from California to Hawaii, where her pale-gray color soon darkened to a deep lead shade with exposure to natural sunlight. Most mornings when he walked by her retirement tank, she greeted him with her usual cheery two-note call. But one day there was no response; she had simply stopped swimming and sunk to the bottom of her tank. "A salute to you, dear old girl," Norris wrote as her epitaph.

VIEILLE THE CHIMPANZEE

Many other individual animals have also been traumatized by captivity. In the Congo, an elderly chimpanzee called, suitably, La Vieille, lived alone for many years in a small concrete cage. She had been part of a chimpanzee colony in a zoo that had eventually been largely abandoned (Linden 1999). The gate to her cage swung open, because the latch had been broken many years before, but La Vieille never left her cramped quarters. She no longer remembered what it was like to be free, and preferred the familiarity of her cage.

PSEUDORCA CATASTROPHE

If an experience is totally new to a species, as happens most often when human activity impinges for the first time into their lives, the result can be catastrophic. The affected animals, frequently marine mammals, have neither a hard-wired behavior pattern nor, often, the ability to respond rationally.

One horrific example involved false killer whales living in the western Pacific Ocean. These animals are often called pseudorcas (*Pseudorca crassidens*), because of a resemblance to orcas (killer whales). Close cousins of pilot whales, they are big jet-black dolphins about 6 m (20 ft) long. In *The Charged Border* (1999), Jim Nollman described a pod of 10 pseudorcas swimming south along the main Japanese island of Honshu, led by an experienced older female. One pod member was a dolphin female who had given birth three weeks earlier to a stillborn fetus that she still kept with her, either tucked under her pectoral fin or bounced along with her snout. She had carted it over 650 km (400 mi) of open ocean, slowing down the pod's progress. When another of them echolocated a school of mackerel swimming 6 m (20 ft) below, the leader gave her "It's time to hunt" signature whistle by forcing air through her pinched blowhole. When she whistled again, three short bursts, the 10 dolphins dove together, coming at the school from 10 different directions. They slapped the water and waved their flukes at the periphery of the ball of fish, confusing and compacting them inward. When the elder pseudorca gave two short whistles, they all emitted sharp bursts of clicks that disoriented the mackerel further, effectively paralyzing those on the edge of the crowded school.

After the experienced female organized the silvery feast, each dolphin in turn, beginning with the youngest, darted into the crowded mass of fish, gulping several down whole before returning to his or her place.

When all were replete, the pod resumed its travel southward, resting now and then on the surface of the water before moving on again. A day and a half later, they arrived at a Japanese fishing ground. Many boats were gathered there to fish for squid during the new moon, the men lowering long hooked lines into the ocean to snag their prey. Pseudorcas each consume about 45 kg (100 lb) of squid a day, so the fishermen noticed when their lines were raided.

The older female called the members of her family together with a distinct pulsed whistle, signifying that there was a large mackerel school swimming nearby. Through the water, she also heard the pulse of diesel engines on boats 16 km (10 mi) away, heading toward them. The boats' owners were angry that dolphins and whales had stolen so many squid from their long fishing lines over the years, but she paid them no heed. For 10 years, the fishermen had organized roundups and killings of bottlenose dolphins and pseudorcas, but the older female was impervious to the danger. She did not lead her pod away from the noisy boats but instead, along with the others, continued to dive deep for food, interact with a few comrades, sing, and even nap.

After the 400 Japanese boats encircled the pseudorcas, each boat lowered a *tsukimbo* into the water—a galvanized pipe, flared at one end, that used a radio signal to produce a loud clanging noise. Together, the devices produced a horrific din that prevented the dolphins from communicating with each other, psychologically incapacitating the animals. The matriarch tried to signal a possible escape route, but none of the others could hear her. For the first time in their lives, they could not use echolocation to make sense of their predicament. The dolphins moved closer together, touching each other, roiling about in alarm, but eventually giving up. They could have escaped by swimming directly under the nets the fishermen set, or tried to break through them, but they did neither. Perhaps the leader wanted the pod to stay together, and knew that the young female pushing her bloated fetus in front of her would not leave her baby. The dolphins had never before experienced such a crisis, and had no response to it.

The pseudorcas allowed themselves to be herded—by boats and netting—into what is known in Japan as the Killing Cove. There they stranded themselves on the beach and slowly died under the hot sun, one by one, over a period of 12 hours. They seemed resigned to death. Pseudorcas have large brains that grow larger with age but, given this pod's unwillingness to at least fight against their entrapment, they did not seem terribly smart from a human standpoint. Their slaughter, however, was not unique. Despite it being illegal, each year up to 23,000 passive dolphins are killed on beaches in Japan; 2300 pilot whales in the Faeroe Islands; and more than 2300 great whales in the open ocean by Norway, Iceland, and Japan together (P. Watson 2006).

Unlike the 10 creatures cruelly killed before their time in Japan, a large wounded male pseudorca near Florida, far away from any harvesting grounds, was given a compassionate death by his pod members (Nollman 1999). The group followed the older male to the shallows, where for two days they took turns buoying his body with their own, so his internal organs would be suspended off the ocean floor and his blowhole would remain above water. His caregivers only left him after he died, slipping away from his body into the open water.

OLDER ANIMALS, LIKE YOUNG ONES, may or may not be adaptive to new situations. The examples in this chapter indicate that many elders are well able to adapt, but this ability may be voided when a species has never before encountered a given situation or is stressed by such things as captivity.

CHAPTER 16

All Passion Spent

FTER A LONG EVENTFUL LIFE, many animals seem happy to wind down their activities, all passion spent. Previous chapters considered two wearied but comfortable pairs of long-lived gorillas, Beethoven and Effie, and Rafiki and Coco (Fossey 1983). The first part of this chapter discusses older animals studied as part of specific population research projects dealing with primate species (Japanese monkeys, Java monkeys, baboons, and chimpanzees). Then it turns to other aging individuals who were observed more casually by field scientists in the course of their research (buffalo, bushbuck, lions, wildebeests, red deer, waterbuck, right whales, and elephants). The chapter concludes with a look at older individuals under the control of human beings (Asian working elephants, a captive tiger, and dogs).

Population Studies

MONKEYS

Japanese monkeys have been thoroughly studied. Japan would seem the obvious place to observe these animals, given their name, but they now also live in Texas. In the 1950s, there were 34 Japanese monkeys in the Arashiyama troop, living wild in the forests near Kyoto. It was difficult for nosy zoologists to see what the monkeys were up to every day, so the men spread sweet potatoes in open areas near the woods to lure them to where their activities could be scrutinized.

Thanks to this extra food, the population not only became less secretive, but quickly expanded, with the monkeys dividing into two daughter groups, A and B. Group A, with about 150 members, began to wander farther afield, raiding local farms and households for food, so irate humans considered shooting them. Their zoological voyeurs were horrified at this possibility. Instead, in 1972 Japanese and American scientists jointly agreed to transfer them to the United States. A new colony was started in Texas, first in a huge fenced enclosure near Laredo, and later in one near Dilley (Fedigan 1991). This colony was named the Arashiyama West troop, to distinguish it from their Arashiyama East relatives still living in Japan.

Some old females remained among the Japanese monkey troop in Japan, and Masayuki Nakamichi from Osaka University decided to observe them, because of their advanced age (1984). Did they behave any differently than their younger sisters? She focused on 14 females, ranging in age from 11 to 29, watching what each individual was doing for 10 hours a day on three different days. Half of these animals were under 22 years old, and half over. During each individual's three observation periods, a researcher noted down her activities, and which companions were near her, on a ten-second-interval checklist. Needless to say, Nakamichi had plenty of data to analyze—over 10,000 observations on the oldest animals, the ones which are of interest here: Op (29 years old), G1 (28), and Op's sister Yu (27). None of these mothers had given birth in the past five years, and none was expected to live more than a few years longer.

Results showed that the older females tended to spend much more time resting than did the younger ones, and that they were more withdrawn from social interactions such as grooming. Grooming is considered to be a tradable commodity that can be exchanged for such things as reciprocal grooming, reduction in aggression, increased access to a scarce resource, or access to a newborn. It is also important for psychological bonding (Nakamichi and Yamada 2007). Fortunately for the three oldsters, their greater apathy did not mean that they lost their dominance and their right to the best food. (Dominance tests were carried out between two monkeys by putting peanuts between them and seeing who grabbed them all.) If a female is dominant when young, she usually remains dominant throughout her lifetime, although she may become subordinate to her own offspring. Op remained dominant to most other females, but subordinate to one of her four daughters and to two granddaughters; G1 was subordinate to three of her five daughters and to a granddaughter. Op and G1 were not close to their elder daughters, but did spend time with their youngest daughters.

Grooming, which strengthened the social bond between groomer and groomee, was the best diagnostic for social activity. Op and G1 were completely lazy when it came to grooming other monkeys; they never did this. They, as well as Yu, seldom even groomed themselves—during a 10 hour period, Op only groomed herself once. Grooming is about parasites as well as social bonding, so the oldsters scratched themselves a lot. Occasionally Op and G1 were lucky enough to be groomed by others, while Yu groomed other individuals more and was also groomed more in return. Unlike Op and G1, Yu spent time with males, sometimes grooming them. In general, because these old monkeys were not involved much with grooming and had no infants of their own, they spent more time alone than did their sociable younger sisters.

Nakamichi (1991) later decided to find out not only what older female monkeys did, but also if where they lived *affected* what they did. The two Arashiyama groups, in Japan and in Texas, were a perfect match for this research. They or their forebears had belonged to the same troop several generations earlier, but now they inhabited different environments. This would be a theoretically clever experiment in genetic versus environmental influences.

As it turned out, there was little variation between the two groups in the behavior of the older monkeys. As in Nakamichi's earlier research, the elder females spent much time resting and largely kept to themselves. Any Japanese monkey troop contains a female and many of her daughters (since it is the males who transfer out of the group when they reach puberty), but in both of Nakamichi's studies, social encounters between young and old adult females were infrequent. When an oldster did interact with another female, it tended to be her youngest daughter far more often than her other daughters—a behavior that would trigger great sibling strife in the human world. If an aging female defended or protected a young relative, this oldster was more likely to be a dominant rather than a subordinate animal, one willing to fight for the high status she bequeathed to her progeny (Hrdy 1981).

More recently, Nakamichi (2003) carried out an age-related grooming study among the 85 adult female Japanese monkeys present in the free-ranging Katsuyama group in Japan. She divided the animals into six sub-groups, two of which contained older animals (age 15 to 22) of either high or low rank. Females of this species prefer to groom related females, but to some extent also groom non-relatives. For both high- and low-ranking females, grooming bouts with unrelated females, but not those with relatives, decreased as the animals grew older. Compared to low-ranking females, the high-ranking females of all ages received grooming more often, and from a larger number of unrelated females. In summary, with increasing age, Japanese monkey females—especially low-ranking ones—are more likely to concentrate their reduced grooming interactions on related females. However, her study also found that older high-ranking females could maintain as great a social attractiveness for grooming as young high-ranking females.

Nakamichi suggested two reasons why older females tended to be more kin-oriented than younger ones in their grooming interactions. First, since grooming one another requires some effort, aging monkeys may choose to use their limited energy on behalf of related animals. Second, since females tend to groom with monkeys of their same age class, older animals will have fewer possible partners as troop members close to their own age die off, and therefore may turn instead to kin. Even though aging matriarchs may cease grooming interactions with unrelated females, they still continue to exchange "coo" sounds with them, and so remain social in that way (Mitani 1986).

In recent research, Nakamichi and Yamada (2007) compared the grooming partners of long-lived female Japanese monkeys in both 1993 and 2003. They found that long-term grooming friendships were common between closely related females, certain unrelated females (such as surviving older partners), and some females closely related to these partners—a web of sisterhood that exists for many older women as well. The monkeys were not close-minded, though, and sometimes dropped a long-time grooming partner, picking up a new one.

Older female Japanese monkeys have special friendships among their troop members, but they do not have a specific social role in their society. From her behavioral study of female Arashiyama West monkeys in Texas, Mary Pavelka (1990) determined that there was no significant difference that distinguished elder females from younger adult females. She watched 40 monkeys—aged 5 to 30 (the oldest)—for a total of 450 hours, and concluded that "it does not appear that the fact of biological aging creates any radically different social situation which would cue the onset of specific role behavior for aged animals."

Working with social groups of Java monkeys, Hans Veenema and colleagues (1997) discovered that old age in this species, especially in combination with a history of low rank, also led to a withdrawal from social interactions with unfamiliar and unrelated animals. In general, older individuals of low dominance became more rigid in their ongoing behavior than did elder animals of high dominance (Veenema et al. 2001). The reduced social interactions of aging low-ranking females with young or unrelated animals may be correlated with the elevated levels of cortisol in socially subordinate monkeys. High levels of cortisol in the bloodstream of humans (as well as baboons) are correlated with stress and with impaired cognitive performance (Beehner et al. 2005). It may be that confronting strangers becomes too stressful compared to dealing with familiar relatives.

Researchers Marc Hauser and Gary Tyrrell (1984) were also interested in reduced social activity in Japanese and stumptail macaques. They came up with three possibilities. First, are aging mothers with older offspring less likely to mix with the female group whose infants play together? This proved not to be so. Second, do older females withdraw from their group because they do not want to waste their energy keeping up social relationships? The researchers could not confirm this. Finally, are older females constrained *because* of their age? How can one prove this?

It does not seem to matter to older female macaques if they are less perky or aggressive than they were when in their prime. They are getting on in years, and do not usually have any youngsters left to raise (although sometimes an older female will care for a motherless young orphan). Most have the time to loaf and savor life. Why not do this? Japanese, rhesus, and Java (long-tailed) macaques also have a social system where dominant females usually remain dominant throughout their lifetimes. Why groom other monkeys if your dominance rank does not require you to? Why get embroiled in the social scene where squabbles drain your energy? Relax. Enjoy.

So with age comes tranquility. Unlike langur monkeys (who grow more belligerent with age), female Japanese monkeys become more serene, a trait especially noticeable in individuals who were particularly aggressive when young. Very old individuals may lose teeth, have wrinkled faces, stiff spines, and stooped shoulders, but they live a peaceful life. Oldsters are less likely to become excited about fights or the presence of nearby predators, and so have a calming effect on troop members. Nakamichi (1991) wrote that helping keep their excitable troop on an even keel "may be one of the most important social functions of old females."

Baboons also slow down over time. Robert Sapolsky, who returned to the same group of Kenyan baboons year after year to conduct further research, was depressed to see his friends slowly age (2001). Naomi gradually became more decrepit than ever, while her middle-aged daughter Rachel appeared tattered, just the way her mother had been. The once-dominant males now had loose, brittle skin and veins that were more difficult to pierce with a needle when taking blood samples.

One bright note was the behavior of Gums—a nondescript male of "vast decrepitude"—and of "ancient Leah," who had so relentlessly plagued Naomi, her social inferior. Leah suddenly left Naomi to her own devices one day, and disappeared from the troop at the same time Gums did. Sapolsky was horrified to lose the two oldest baboons on the same day. What could have happened to them? Could they have been eaten by hyenas or leopards at the same time? Ten days passed. Then Sapolsky happened to drive back to camp by a circuitous route through an area that his troop almost never visited. And there he saw Gums and Leah sitting together. They both gave a start at the sight of the vehicle, although Leah, at least, had with equanimity accepted people nosing about her troop for nearly 20 years. In alarm, she loped stiffly into a nearby thicket, followed by the equally panic-stricken Gums. The next day Sapolsky drove by the same area again and, in the distance, detected the two friends resting together on a windswept ridge in a field of wildflowers. He never saw them after that.

The baboon Athena, a menopausal female more than 20 years old, did not have a male friend. Like Gums and Leah, she disappeared from the troop for weeks at a time, but apparently alone, and finally was not seen again (Smuts 1985).

As extreme old age creeps up on a baboon, life becomes fragile. A male might wait a day or two for his ailing female to keep up, but no one will wait for a decrepit male who lags behind the troop as it forages for food. One feeble old hamadryas male arrived at the sleeping cliff two hours after the rest of the troop, when it was growing dark (Kummer 1995). He hauled himself slowly up to a sleeping place, where he was greeted in a friendly way by other members of the troop.

The elderly female Narba was equally late on another occasion. After the troop had settled down on the cliff for the night without her, her male friend, called Friend, got up, stared intently to the north, then climbed down the cliff and set off in that direction. After about 10 minutes he returned, along with Narba.

Many old baboons, both male and female, display far more risky behavior by being out alone than they would have when younger. Narba later stayed away for a week, very vulnerable to predators, before again visiting her family. After that, she disappeared and was not seen again. Kummer (1995) reported that very old lone baboons will not suffer for long; they last for only a few days before dying or being swiftly killed by a hyena or leopard.

CHIMPANZEES

In the Mahale Mountains on Lake Tanganyika, researchers have been studying chimpanzees since the 1960s. Here, as in Gombe, there were few old individuals of either sex, all of whom remained social by keeping in contact with at least one other chimpanzee (Nishida et al. 1990). Wanaguma, in her late forties, was one of the oldest females in the Mahale M group, incapacitated to some extent because she had a cataract in one eye and could not see well (Huffman 1990). She had thin hair, worn teeth, and a brownish rather than a black coat. (When they are very old, chimps become emaciated and bony, with little hair on their heads and shoulders.) Like all the older chimps, now Wanaguma did not travel as far each day, climb as often into fruiting trees, or spend as much time in grooming sessions and other social activities as the younger animals. She was no longer reproducing, but fortunate to still have a son in the group, NT, who was a prime alpha male. She sometimes hung out with him and, because he belonged to the core adult male subgroup, with these males, too. Other chimpanzees no longer attacked her, and sometimes even groomed her, although, with the exception of her son, she almost never groomed them back. If somebody caught a small animal to eat, they were likely to give her and the other older chimps some of the meat, while ignoring hungry younger animals.[1] Wanaguma was given respect because of her age—a cushy, restful life at long last.

Wankuma was more social with the other females than Wanaguma, perhaps because she had no other kin in the group; her daughter MM had recently died. She tended to socialize with two young orphans, feeding and traveling together with them, although she also spent time with the seventh-ranked male. Unlike Wanaguma, this older chimpanzee groomed other animals about as much as they groomed her. Michael Huffman (1990) noted that "the absence of adult offspring to depend upon and receive attention from may have led her to make a more concerted effort towards associating with others."

Kagimimi, one of the oldest males in the M group, was also less sociable (judging by the time he spent in fairly close company with others) than he would have been when younger, but he was still twice as sociable as Wanaguma. He kept limited but stable contact with the adult male core subgroup, and also hung out at times with older females and with young males. As with Wanaguma, he was far more likely to be groomed by others than to groom them. Older males sometimes threaten to attack prime males, but these younger males do not retaliate, wanting to keep the peace if possible, and perhaps realizing that insolence from elder chimps would not destabilize established social relationships.

Casual Observations

BUFFALO, BUSHBUCK, AND WILDEBEEST

Dian Fossey (1983) was pleased to have buffalo (*Syncerus caffer*) living around Karisoke, her mountain camp. One couple was immensely old, the body of the male crisscrossed with scars, probably from encounters with poachers, traps, and other buffalo during his long life. The boss at the base of his horns was heavily worn down, and the horns themselves were shattered nubbins, shadows of their former selves. Fossey named the bull Mzee, meaning "old man" in Swahili. When Fossey first encountered Mzee, he was accompanied by an equally elderly female. This was rare, because usually female buffalo remain in their herds when they grow old, while males become solitary. This female tended to precede him as they moved about on the mountain, as if aware of his failing eyesight. In time the female disappeared, and Mzee browsed alone.

It was surely the presence of Dian and her trackers on the mountainside that offered protection to some wild animals by keeping poachers at bay. This would explain the long life of the buffalo, as well as that of the magnificent old bushbuck, Prime II, who ranged near her cabin. He was so old that, according to Fossey, he could "barely walk/wobble" (quoted in Mowat 1988), yet he was the adult male in a group of two adult females, two 2-year-olds, and two yearlings. After Dian had been murdered, her good friend Rosamund Carr, broken-hearted, attended the funeral at Fossey's camp (1990). When Carr broke away from the ceremony to grieve in a nearby meadow, she saw Prime II, whom Dian had often admired from her window. "He had always been wary of people and noise, but now—despite the crowd of strangers and commotion nearby—he was standing on the lawn gazing toward Dian's house, as if to say good bye."

A group of older male buffalo also hung around the wilderness camp set up by Mark and Delia Owens in Zambia's North Luangwa National Park, as they related in their book *Secrets of the Savanna* (2006). When the Owens first arrived in Luangwa in 1986, the buffalo were extremely wary, because so many thousands had been massacred by poachers. For years the Owens, using binoculars, could glimpse buffalo in the thickets on the far side of the river, sometimes peering suspiciously back at the human intruders. Slowly, other animals began to make themselves at home in the camp. The buffalo would glimpse puku antelope, warthogs, and elephants among the huts that now formed a sanctuary where these animals were protected from lions as well as poachers. Gradually, the buffalo felt safe enough to join the others.

By 1990, each morning these massive animals slowly wandered into the settlement from the long grasses by the riverbank, where they had spent the night. They were no longer interested in sex, and in any case were too tired and slow to keep up with their herds. At first they came at night; once, after dark, Delia nearly walked into a buffalo grazing along the path to the kitchen. He gave a resounding deep growl of surprise, but fortunately did not chase her as she fled to her cabin. They named him Brutus, and he was soon joined by Bad-Ass, an equally large male who had the disconcerting habit of lowering his head and hooking the air with his horns. Each day the two beasts stood near the camp, staring at the people walking about from hut to hut. Later Stubby (with only a stub of a tail) and Nubbin (with worn-down horns) joined them. The four buffalo were now called the Kakule Club, the ChiBembe word for elderly male buffalo. Eventually they felt at home enough to lounge about near the kitchen, chewing their cud, grazing in a desultory manner, or sleeping while oxpeckers scouted along their hefty bodies, hunting for ticks. The Owens had mixed emotions about these newcomers: they were unnerved to have such potentially dangerous animals so close at hand, but pleased that after so many years of their tribe being slaughtered, the buffalo finally felt comfortable among people.

Lone older buffalo in the Congo's Albert National Park also changed as they aged—from the most dangerous animals in Africa to beasts behaving in a familiar way, like cattle in the Swiss Alps. They were no longer hunted for their horns or harassed by their peers, so they had become quite soft-hearted. Heini Hediger (1968) was so astonished by their tameness that he measured their flight distance (the nearest that people could approach them on foot without them fleeing) to be the "seemingly ridiculous figure" of 12 m (40 ft).

Most of the older males that were mentioned as loners likely had reduced testosterone in their blood, so that they were not very interested in sex. They probably missed companionship, though. The Owens described older lions in the Kalahari Desert who were no longer strong enough to hold a pride of lionesses together (and probably were not even interested in doing so). The aging warriors would stand on a termite mound, "looking far across the landscape for minutes at a time, 'cooing' into the wind, inviting social contact among pride-mates who were no longer there to hear them." Sometimes these males would encounter other older males, even former enemies, and hang out with them, "just to be with someone, anyone, rather than spend one more minute alone."

A lone bull wildebeest in the Kalahari would follow the Owens's Land Rover whenever it came near him. He seemed to want company, even if it was only a vehicle. "Individuals of a social species will do almost anything to avoid being alone in an empty desert—or in a river valley full of lions and poachers. Far away from their relatives for most of their lives, many male mammals are at first aggressive, adventuresome competitors, then finally find themselves alone and, perhaps, lonely."

WATERBUCK

Animals eaten by others usually do not live long enough to die of old age—as they slow down, they are likely to be taken by predators—but a few become loners. C. A. Spinage (1982) extensively studied waterbuck (*Kobus ellipsiprymnus*) in Uganda. They seldom lived more than 10 or 12 years, but one older female (judging from the cementum line count of a tooth root) survived to be 18, probably because she had exceptionally hard teeth—as with many species, at best waterbuck last only as long as their teeth do. This female was a gaunt soul, with many gray hairs and such stiff joints that a veterinarian surmised that she would not survive the rainy season. However, she did, recovering from her stiffness (despite the rain) and even producing a calf. As female waterbuck grow older, they still visit males if they feel like it, but they gradually restrict their everyday range to increasingly smaller areas where the prime females, who may harass them, do not go.

Waterbuck often live in large groups of up to 70 animals. During the breeding season, the pugnacious prime males choose and defend a large territory by fighting other males. There, they mate with females who visit their plots of land. Older, 10 or 11-year-old males have smaller territories, usually located far from the desirable waterfront areas, than the younger males. But this does not mean that they have lost their territorial urge. Spinage reported that male Y9, when he was at least 10 years old, tried three times (that Spinage knew of) to retake the prime piece of real estate that he had once considered his own. Three times he was driven from the area, on the last occasion streaking past Spinage with the new owner in hot pursuit until they reached the territory's boundary.

RED DEER

In his book *A Herd of Red Deer* (1969), Fraser Darling reported that old red deer stags, like solitary elder males of many species, also become antisocial, having more or less given up on life. In Scotland, such a male, incisors worn to nothing, would lie low in bracken during the daytime to escape detection and biting flies, and then raid cornfields at night. His only movements would be between his sleeping place and his eating place. His antlers would be poor, his gonads would show no seasonal enlargement, and he would have no interest in females. "What does he do? Nothing, for the most part. He eats and sleeps and looks morose." He frequently would meet his end by a bullet in the moonlight.

Right whales, as they age, also become more somnolent than their younger relatives. Troff, the aging mother of many calves, liked to lie motionless for hours on windy days, like a submerged reef, below an observation hut of a whale research station in Argentina (R. Payne 1995). When she slept, she rose slowly now and then to the surface (as if in a dream), took a breath, and then slowly sank again to the bottom. Right whales are so fat that they float when completely relaxed, and their rate of breathing drops drastically (Ackerman 1998). Troff was obviously in almost complete repose for most of the day.

AFRICAN ELEPHANTS

Elephants are so social by nature that it is hard for them to be alone. Yet the very oldest elephants *are* loners in that they can no longer keep up with their herd as it travels about foraging for food and visiting sources of water. They no longer seem to be interested in sex. Nor can they consume food easily with their worn teeth. They never lie down, because they would be unable to rise again (Crandall 1966).

Lone females are rare, though. In their 1966–67 campaign to exterminate 2000 elephants in Uganda, a culling team discovered only six solitary females (Laws et al. 1975). These former matriarchs, age 52 to 59, moved slowly, looked elderly, and were in poor condition; they were neither pregnant nor lactating. "It is likely that they had not much longer to live," the culling team wrote, a prediction they made come true by shooting them all.

Katy Payne (1998) and her Zimbabwe colleagues put a radio collar on Jabula, an old female, so they could plot where she wandered from day to day. After taking 413 separate readings on her position, they worked out that she was almost always within a small (23 sq km) area.

Lone male elephants are more common and more aggressive than lone females, perhaps from one or more of the following causes: the ignominy of losing their place as an alpha male to a younger bull; sexual frustration; or a probable decline in vigor brought on by such factors as injury, cardiac problems, bile stones, various diseases, and intestinal parasites (such as roundworms, hookworms, and tapeworms) that cause anemia and reduce protein uptake (Sikes 1971; Chadwick 1992). Some large bulls have a long history of being hunted for their tusks, so they are clever, very dangerous, and difficult to shoot. They become rogues and marauders who will not hesitate to attack a person or a vehicle, or to feed on planted crops.

The most elderly elephants of both sexes gravitate to places where they are well hidden and can survive without having to move far—often in thick cover near a river or swamp, where the vegetation is soft and easy to chew with their worn teeth (Buss 1990). When these loners die, their flesh is eaten by scavengers and their bones and tusks scattered, gnawed by rodents, and soon concealed by vegetation. Could this be a proto-typical cemetery?

Since the remains of so many elephants who die naturally are never found, there is controversy about whether or not there might be an elephant graveyard where old elephants go to die, surely triggered by the wishful thinking of hunters who would dearly love to collect their huge tusks for ivory. Certainly some older bulls, sometimes together with younger accompanying males, do congregate in places such as the lower Tangi River region in Uganda. This area is difficult for hunters to access, has green vegetation growing along the river's edge even in the dry season, and is a permanent source of water for drinking and bathing (Buss 1990). One ancient animal, widely known as the Tangi Bull, stood 3.1 m (10.2 ft) at his shoulder, with large sunken hollows above his eyes and a protruding spinal column covered by wrinkled skin. After he died of natural causes, his left tusk was found to weigh 72 kg (158 lb), and the right 66 kg (145 lb). One can only imagine that these enormous structures were a gross handicap in his last years, when the decrepit bull was no longer using them to fight or knock over trees.

The future of African elephants who remain outside of national parks in Africa is not bright, with burgeoning human populations encroaching on land where these elephants have always lived. The elephant population in Knysna Forest in South Africa, for example, has been decimated. Some individuals were shot in the 1920s, and others died subsequently. Biologist Lyall Watson (2002) reported that in 2000, only one elephant, an old matriarch in her fifties, remained in the forest. It hurt him to think of a solitary female. Under natural conditions, females are so social that they are never, ever alone during their entire lives. They are always surrounded by mothers, aunts, sisters, youngsters, and infants.

Watson saw this lonely soul on the shore of the Indian Ocean, looking out to sea. He felt that she had come there because there was no longer anyone she could talk to in the forest. Near the water, she could at least find peace in a spot where the rumble of the ocean surf gave off soothing infrasounds too low for the human ear to hear.

But then Watson became aware of a throbbing in the air, a sensation so slight that he might have missed it if he had not been attuned to infrasounds sent out by elephants to communicate with each other. He glanced out to sea and there, 100 m away, was a blue whale floating at the surface, pointing toward the elephant, her blowhole plainly visible. These whales also use infrasound to send messages to each other, in their case often many hundreds of kilometers through cold ocean waters. The largest mammal on earth and the largest mammal in the ocean seemed to be communicating! He imagined that they were sharing thoughts—matriarch to matriarch—about being old and alone, about the progeny they had borne, about their past social lives. He wrote that "I turned, blinking away the tears, and left them to it. This was no place for a mere man."

Old Animals under Human Control

ASIAN ELEPHANTS

Douglas Chadwick (1992) visited the Mudumalai Sanctuary Elephant Camp, situated within an Indian reserve where domesticated elephants and their mahouts help with selective teak logging. These elephants, which do far less harm to the environment than bulldozers, have a work life of about three decades, far longer than a machine. Retired elephants, called pensioners, lead a pleasant existence here, enjoying a daily bath in the river and a scrub by their mahout to keep their skin clean and free of parasites. Their food includes rough wheat, molasses, rice, and coconut.

One female, Tara, after retiring from work in her mid-sixties, then got pregnant, a decade after most females cease breeding. She lived on in the camp until she was 78. Chadwick fed wheat cakes, one at a time, to another retired cow, 73-year-old Godavri. He was also introduced to a female, Rathi (the Hindu equivalent of Aphrodite), who was 58 years old. Rathi had had 10 calves, all sired by wild elephants outside the camp because she would not accept camp tuskers; outbreeding was an accepted practice to prevent inbreeding among the "tame" local animals.

The day before Chadwick arrived, the mahouts had given a party for the elephants, to celebrate the end of the long logging season and thank them for their help. The mahouts had painted their charges' foreheads in festive colors, held a ceremony to praise their strength and abilities, and then given them a feast of rice, fruits, and sweets that the men had purchased with their own money.

It is heartening that in the United States, there is pressure from groups such as PETA (People for the Ethical Treatment of Animals) to encourage rural sanctuaries such as the one in Tennessee, where older elephants can retire from circuses and zoos and end their lives in peace. Most of these are female, because adult males are too strong to be kept safely in captivity. For this reason, a 1952 survey indicated that of the 264 elephants in the country, only 6 were males (Alexander 2000). Today, the average zoo elephant dies in her early forties, with her lifespan greatly cut short by her captive condition (Gorman 2006).

A CIRCUS TIGER

One old circus tiger became more relaxed as the years passed (Thomas 1994). Rowena worked for trainer Klaus Blaszak, who had a traveling circus show that included a number of elderly tigers. Rowena enjoyed the acts but, with age, became tired by the time the last performance began at 11:00 p.m. She would be sleeping in her cage when the music announcing the tiger routine was played. Immediately she would rouse up, pad willingly into the ring, and jump onto her stool. Then she would begin to snooze again. "Her eyelids would close, her head would droop, her thighs would loosen, and her knees would slowly spread. At last, even her tail would lose its tension and would hang straight down." She managed to execute her part in the act—she was a veteran performer—but when it was finished she was glad to trot back to her cage, where she slept soundly until morning.

DOGS

As a dog approaches the end of life, his or her brain decreases in size, and hearing, eyesight, sense of taste, and sense of smell are all less keen. Compared to a young dog, he or she is more likely to forget things, be harder to train, and be slower and less responsive to commands. However, as Stanley Coren described in his book *The Intelligence of Dogs* (1994), such oldsters can be roused from inertia. He reported that a rescued cairn terrier called Whistler began obedience training at the age of 11, and at 12 had earned his obedience diploma. "Whistler walked out of the ring with his tail beating as quickly as any proud puppy, and if his master had a tail, I'm sure that it would have been wagging just as happily."

Coren also described the remarkable story of Shotgun, a brown Labrador retriever, which illustrates not only an old dog rising to the occasion, but also his intelligence and loyalty. Shotgun was 11, and could no longer do such things as jumping onto the sofa. He smelled smoke one night, and began barking furiously to awaken the household. When this did not work, he moved as quickly as his arthritic legs would let him into the parents' room, and managed to clamber onto their bed, barking until they responded. The parents rescued their two young sons, but thought that their eldest child, Melissa, would have already escaped, awakened by the commotion. She had not. Shotgun lumbered into her smoke-filled room to find her standing by her bed, confused and crying. Shotgun grabbed the sleeve of her nightgown and dragged her forward. When he realized that escape via the front door was barred by the flames, he led her to the back door instead. By standing on his hind legs, he was able to push up the screen door latch—cutting his nose in the process—and lead Melissa into the backyard. There he let her go, and began to lick his singed paws.

Coren noted that although Shotgun was old, and thus slower and less reliable than he had been, this did not mean he was stupid. He had sensed danger, warned the household, realized that one child had not escaped, and rescued her by working out how this could be done in a crisis situation. All five members of the family, his pack, owed their lives to this old dog.

AT THE END OF THEIR LIVES, most animals are no longer interested in leadership, reproduction, fighting, or even much social life; if they do socialize, it is likely to be with their relatives. Many become loners even though they may sometimes wish for company. Although they limit their movements to small areas where enough food and water is available, these oldsters at last have time to rest and relax. They do not necessarily lose their spark, though: waterbuck Y9 was still desperate to regain his lost territory, and the arthritic dog Shotgun was able to rescue his human family from fire. Death, however, is inevitable. The next chapter deals with the natural dying and death of old animals in the wild, in captivity, and as the companion animals of people.

CHAPTER 17

The Inevitable End

NIMALS IN THE WILD RARELY DIE a slow and agonizing death, as people sometimes do. There are predators around to snap up those who are too slow to escape. If they avoid predation, there is still no medicine to stave off disease. When their teeth are worn down, they can no longer eat efficiently. Oldsters are usually alone or with a friend—they are no longer able to keep up with their group (such as elephants), or have been driven out of it (such as hyenas). They are often gaunt, with less fur or hair than in their prime, and that which remains may be gray or white. They may suffer from obvious ailments such as arthritis and ectoparasites, or from unseen diseases. This chapter contains some examples of the final days of some old animals, followed by a discussion of the rites of death practiced by elephants and gorillas (both in the wild and in captivity), and the rites people have for their old animal companions.

THREE UNGULATES

When B-ram, a mountain sheep, died at nearly 13 years old, Valerius Geist (1971) was able to document his last months. Despite his age, evidenced by a slightly sagging belly and back, B-ram successfully mated with females on Ghost Mountain in his last rutting season, because he had the largest horns of all the males in the area. By the end of the rut, though, he was having trouble running as fast as the ewes and the other rams. In January, he joined the group of males who bedded down each winter night in a cave on Sanctuary Mountain. He spent some time each day resting like the other males, but more time foraging for food. He was now limping, holding up his left hind leg when he moved, which made it difficult for him to balance while pawing through the snow to reach the edible plants beneath.

By early February, B-ram looked gaunt. He still rested in the communal cave at night, but during the day he stayed nearby, searching for food while the other males wandered farther off. On several occasions, 6-year-old G-ram butted and pestered him, as subordinates sometimes do to sick or injured dominants—perhaps seeking revenge for past harassment? In late February, B-ram abandoned the cave, instead bedding down in the open, subject to cold and arctic winds, moving lower and lower on the mountainside each day. On March 12, Geist found his body, with tooth marks around his throat where he had been attacked by wolves.

In their prime, male African buffalo live in large mixed herds, but when they are older than about 12, they tend to subsist alone or in small groups. Mzee was the ancient buffalo who roamed with an equally aged female near Dian Fossey's mountain camp. After the female died, Mzee carried on alone. If he heard Fossey's voice in camp in the early evening, he would slowly feed toward her, as if wanting her company. Eventually he let her scratch his withered rump. Early one morning, a woodsman discovered him lying dead in a serene grassy hollow below two mountain peaks. Although he had lived all his dignified life under threat of poachers, he had finally defied them in death (Fossey 1983).

The eminent naturalist Sigurd Olson (1987) described the death of an old white-tailed buck (*Odocoileus virginianus*) in the Quetico–Lake Superior area. He had been pursued by a pack of wolves, chased to an exhausted standstill, and then swiftly brought down by the predators. "He might have died slowly of starvation or disease, but he died as he should when his time had come, fighting for his life against his age-old enemies, dying like the valiant warrior he was out on the open ice."

ELEPHANTS

Elephants live only as long as their teeth are functional. These comprise six heavy, laminated molars in each half jaw, which slowly develop during the life of the animal and move forward in sequence, like a glacially slow conveyor belt; only one or two molars in each half jaw are available for chewing at any one time. After the sixth and final teeth have moved forward into place and become worn down, the animal can no longer chew food adequately. This is why very old elephants tend to haunt river areas, where there is soft vegetation.

Is there an elephant graveyard where individuals go to die? Ivory hunters would love to find such a place, especially if it contained the remains of old animals, who have the biggest tusks. Sometimes large collections of elephant bones are indeed found in one place, and there are various stories to explain such accumulations (Buss 1990). One hunter came across many skeletons east of Lake Chad; he believed that the animals, migrating to this region from a terrible drought, died after drinking water from poisonous natron springs.

Another story is told of Africans lighting a huge fire around an immense herd of elephants, and then making sure they died in the conflagration. Hundreds of carcasses littered the ground, many of them of older animals.

Louis Leakey postulated that Africans living beside trade routes may have stashed tusks near their homes, which were later found and thought to mark the place of death. It was assumed that the tusks survived while the rest of the bones had rotted away (Buss 1990). Or, scattered elephant bones may be the aftermath of a massacre by poachers.

No matter what the cause of death, once the flesh has been removed, the bones of an elephant, but not the bones of other species, remain of interest to living elephants. Most probably happen upon such bones now and then. They examine these, milling around them, perhaps picking them up in their trunks and smelling them, perhaps moving them respectfully and covering them with grass or earth. Cynthia Moss (1988) once brought the jawbone of a dead female to her camp, so she could age the animal by examining her teeth. A few weeks later, when the family of that female passed through the camp area, it made a detour to examine this jaw. The female's 7-year-old calf remained with the large bone for some time, turning it over with his feet and trunk, as if remembering his mother.

WHALES

Roger Payne does not know the expected life-span of the right whale, because he had only observed this species for about 24 years. There is some evidence, though, including a photograph of a female taken in 1935, when her calf was killed (the last right whale slaughtered by whalers in North America before such killing was banned). At that time, she must have been at least 5 years old to be a mother. Over the years she had been photographed four times, including in 1992. She still did not look elderly then, but she must have been at least 62 (R. Payne 1995).

How do right whales die, since they have few natural predators (besides human beings), and since harvesting them has been banned (Clapham 2004)? Sometimes they become entangled in nets set to catch fish—an alarming 10 to 20 percent of all individuals become trapped this way in any one year. Right whales are powerful enough to drag fishing gear away with them (unlike dolphins, who usually drown), but they may never be able to shed the gear. Some die because it prevents them from swimming and feeding normally, and others from infection caused by the rope becoming embedded in their flesh. People sometimes try to free a right whale from such entanglements, but they are usually too apprehensive about their own safety to be successful; in their panic, whales may thrash their huge tails and drag a boat for kilometers across the water.

Other right whales die because a ship runs into them while they are resting on the water's surface. (Right whales were given the name "right" because they had the proper attributes to make them ideal fodder for whalers —they were slow, had lots of blubber to make them float easily, and spent much of their time on the surface of oceans.) The body of one very elderly female, Staccato, who had borne at least six calves, washed up dead on a Cape Cod beach in 1999. She had first been seen in 1977, and recognized again over the years by the callosities (hardened skin bumps) and barnacles spread over her large head. At first biologists could find nothing wrong with her, but when they examined her right lower jaw, they found it was massively fractured. After a collision with a ship, she had remained alive for about a week before dying from her injuries.

Whales also die naturally of old age, as Shari Bondy observed during the winter of 1999 in Baja California, Mexico. She saw 10 elderly gray whales clustered together in the shallow water offshore; this was unusual, as normally they would be spending their time with the mothers and calves (D. Russell 2001). The following day, one of these older whales died. "It made you wonder if they had not been paying their last respects," Bondy said.

Like all other cetaceans, aging sperm whales have to keep one portion of their brains perpetually alert, night and day, in order to make sure that they keep breathing by surfacing regularly. As a deep-water species that likes to have at least a kilometer of ocean beneath it, sperm whales do not have the option right whales such as Troff had—she rested for hours in shallow water, merely extending her head upward to breathe. Older sperm whales carry on with their nomadic lives for as long as they are able. When they are very old and near death, individuals sometimes strand themselves on a shallow beach, where they can die without drowning (Whitehead 2003).

COMPANION ANIMALS

The deaths of companion animals are usually not as swift as those of wild animals. The last three older dogs in the pack studied by Elizabeth Thomas (1993) started to die off at about the same time. The remaining alpha male, the husky Suessi, seemed to develop something like Alzheimer's disease. He almost forgot that human beings existed, although he remembered his sisters and the fields nearby. When his arthritis became so severe that he could no longer stand up, Thomas carried him into the veterinarian's office to be euthanized. When she returned home, she showed the other dogs his collar. They stared at it calmly; then, after slowly smelling her clothes and her hands, they looked at her quietly, "as if thinking things over, or taking things in."

Inookshook died a few weeks later, also unconscious and at peace. Fatima lived on alone. Her diabetes worsened and, learning that her insulin injections made her feel better, she nudged Thomas for them, just as she used to nudge her for food. One day Fatima left the house through the dog door and walked slowly off into the woods. She never came back. Thomas and her husband hunted for her everywhere, but they found no trace of her, not even her collar. Fatima felt ready to die, and had gone off to be alone.

James Thurber agreed that dogs often go away by themselves at the end: "The last ineluctable scent on a fearsome trail, but they like to face it alone, going out into the woods, among the leaves . . . enduring without sentimental human distraction the Last Loneliness, which they are wise enough to know cannot be shared by anyone" (quoted in Masson 1997).

Rarely, an animal becomes noteworthy not because it died, but because its wealthy human companion did. Billionaire Leona Helmsley left $12 million in a pet trust for her white Maltese dog, Trouble (Pearce 2007);

Helmsley's brother, who would care for the dog and administer the fund, received $10 million from her will.

ANIMAL PAIRS

Pairs of animals who have been together for a long time often mourn at length when one of them dies. This is true for parrots, both in captivity and in the wild, who have mated for life. The remaining half of the pair sometimes isolates itself from its flock and neglects to preen its feathers, undermining its ability to fly efficiently (Burger 2001). "Such lost, forlorn birds can be seen just standing around, barely feeding themselves, letting down their guard, not doing the things they need to do to keep themselves healthy."

In her book *The Emotional Lives of Animals*, Betsy Webb (2007) described the long friendship between the llamas Boone and Bridger, who lived with her family first in Colorado and then in Alaska. In Alaska, the two Bs had become friendly with two other llamas living there. When he was 27 years old, Boone suddenly lay down on his side and, too weak to get up again, died. The next day his life partner, Bridger, died in the same fashion, lying next to his body. After these two were buried in a nearby field, the remaining female llama stood and stared at the burial spot for the best part of two days, while the male stayed in the barn and wailed. On the third day, both llamas resumed their normal activities.

At the zoo in Lucknow, India, an old elephant also died of grief for her friend (*Record* 1999). Damini, 72 years old, had lived alone in the zoo for five months before she was joined by a younger pregnant female, Champakali. The two elephants took to each other instantly. After seven months, though, Champakali died giving birth to a stillborn calf. Damini was so upset that she stopped eating. When her legs would no longer support her, she lay down on her side, head and ears drooping, trunk curled. She refused all food and drink and stopped moving, no matter how the zookeepers tried to please her: tempting her with sugarcane, bananas, and grass, her favorite foods; erecting a tent over her strewn with fragrant medicinal grass; and using a water spray and fans to cool her body. Damini died 24 days later, covered with a mass of bed sores, the loose skin hanging from her bones.

The most poignant death of any individual animal was surely that of Martha, the last passenger pigeon (*Ectopistes migratorius*) (Schorger 1955). This species slowly became extinct, with the once vast flocks that darkened the sky reduced to seven birds in 1908. In 1910, there were two birds left. After that, there was only Martha from Wisconsin (named after Martha Washington), captive in the Cincinnati Zoo. On September 1, 1914, when she was anywhere from 17 to 29 years old (experts varied in their estimations), she was found dead on the ground. Her body was frozen and sent to Washington, DC, where it was examined, stuffed, and mounted in a glass case in the Smithsonian Institution's National Museum of Natural History.

Rites of Passage

IN THE WILD

Highly intelligent animals have rites connected with the death of their fellows, such as those elephants have when they come across elephant bones (see earlier in this chapter). Occasionally an elephant will pay homage to a dead comrade by standing still and facing away from the body, reaching back now and then to touch it gently with a hind foot. It is as if the mourner does not want to look at his or her dead friend, but cannot bear to leave the body, either (Moss 1975).

Peter Jackson (2007) reported coming across the body of a baby elephant in Botswana's Chobe National Park. The youngster had been killed at night by lions and partially eaten. While he watched, a matriarch with tattered ears, indicating her great age, led columns of about 100 elephants to the small corpse. They crowded around the bloody remains, some stamping their feet and snorting in the direction of the lion pride that lay nearby. Most, however, lightly touched and sniffed the little body with their trunks, then moved a respectful distance away, standing in silent groups. Eventually, the matriarch turned and slowly led the herd back along the valley, the way they had come.

Elephants sometimes bury dead carcasses. An extreme example occurred during an extensive cropping (killing) operation in Uganda, when scientists and park officials collected the ears and feet of slaughtered elephants to sell later for the production of handbags and umbrella stands (Moss 1975). They stored these body parts in a shed. One night, presumably smelling the objects from a distance, elephants broke into the shed and buried the ears and feet, to the moral discomfiture of the men.

Elephants also, in effect, buried the carcass of a foolhardy lion who had leapt onto an elephant's shoulder, not realizing the risk involved (K. Payne 1998). Using her trunk, the elephant immediately grabbed her attacker by the tail and slammed him into the ground again and again until he was dead. Then the herd covered his body with branches they broke from nearby bushes.

Gorillas may also bury their dead, as Cindy Engel reported in her book *Wild Health* (2002). Gorillas have been observed covering a gorilla corpse with leaves and loose earth collected nearby, and gorilla bodies have been found under loose vegetation in the forests of Uganda. Perhaps this is why Dian Fossey was unable to find the bodies of some of "her" gorillas who died, no matter how hard she and her helpers hunted for them. Maybe such burial activity is useful in limiting infection and disease during putrefaction? Or in lessening the likelihood that the body will be found by predators? We just do not know.

IN CAPTIVITY

At least one zoo has made special arrangements for older animals who are dying. In the 1980s, officials at the Chhatbir Zoo in Chandigarh, India, had the less-than-brilliant idea of crossbreeding Asian and African lions in order to attract more visitors to the zoo (Kumar 2006), as Asian lions had been nearly exterminated by maharajas and princes hunting them as trophies. Unfortunately, although 70 to 80 hybrid lions were produced, these animals were frail, with uncertain immune systems and weak hind legs. The zoo officials have now created an "old age home" for lions too feeble to defend themselves. This site, away from the main exhibit area, provides these lions with "all the facilities to live a happy life in their last years." They are sometimes fed boneless meat, and are given good medical care. "Even if they are meant to die, it does not mean we kill them by not treating them," a spokesman said.

As early as 1994, an older female gorilla who had died of cancer at the Boston Zoo was given a wake (Bekoff 2005). Her body was laid out so that her long-time mate could visit it. "He was howling and banging his chest . . . and he picked up a piece of her favorite food—celery—and put it in her hand and tried to get her to wake up." The human observer, Donna Fernandes (later president of the Buffalo Zoo), found the scene so emotional that she wept. Subsequent wakes have been held for other gorillas: Omega, a male at the Buffalo Zoo, and Babs, a female who died at the Chicago Zoo (Brookfield Zoo). As a television program reported, Babs was laid out in a room, and members of her family quietly filed in one by one, each approaching their leader and gently sniffing her body.

Increasingly, cutting-edge zoos are allowing rites to be played out when one of their older animals dies. In 2006, when the 46-year-old elephant Lucy had to be euthanized at the Milwaukee County Zoo because of illness, Brittany, the other elephant at the zoo, was allowed into Lucy's stall so she could grieve by the body of her friend (*Record* 2006). Similarly, when Patsy, a 40-year-old African elephant at the Toronto Zoo, was "put down" because of excessive pain from arthritis, the other six elephants were allowed to spend the night following her death with Patsy's body, so they could "go through a mourning process" (*National Post* 2006).

The Magnetic Hill Zoo in Moncton, New Brunswick, Canada, planned a winter celebration and open house for their Siberian tiger, Tomar, who had lived there for nearly 20 years (*Record* 2007). The beloved elderly animal was dying of kidney failure, and would soon be dead. He had been alone since his mate, Pasha, died three years earlier from eating donated meat from an animal who had been euthanized with barbiturates.

COMPANION ANIMALS

People, too, can have ceremonies to help them cope psychologically with the death of a beloved old companion. Rather than coldly sending an aging animal off to the vet for him or her to be "put down," they can hold a small ritual of remembrance and compassion. As Linda Tellington-Jones (quoted in Fleetwood 1998) described it, first a person must ask the animal who is becoming decrepit or progressively sicker if he or she wants to be released from their suffering. This may sound silly, but she insisted that many people who have never really talked with their animals now ask this question, and some feel the animal gives them clear permission to let them die. She felt that many pets die soon after this, because they have been holding on to life long after they wished to leave it; they were fighting against their ailments because they knew their human companions could not bear to let them go.

Once an animal has given permission or is deemed to be in too much pain to carry on, his or her human companion calls in a few friends who know the cat, dog, or horse. They sit quietly around the animal, gently grooming and stroking him or her. They talk about what a wonderful life the pet has had, about how much joy the pet has given to human friends, about the sorrow they feel on this sad occasion, and about how it is now time for the animal to move to another stage in the cycle of life. Then the pet's caretaker holds the animal close while he or she is injected and makes the passage into stillness and freedom from pain. In our culture, we are often taught not to make an open show of grief, but what rational sense does this make if our hearts are breaking?

If such a rite is difficult to arrange privately, pet funeral services are now available in some cities (Davidson 2005). These businesses can pick up the dead pet and arrange a viewing, a eulogy, burial or cremation, a monument, and grief support. "People feel a need to have some kind of ritual that attends to the death of the animal so that it is not simply thrown away, like a thing would be."

Probably no dying animal has had a more unusual directive than Pat, the brown Irish terrier who lived with the Canadian politician Mackenzie King. After the deaths of his mother, brother, sister, and two dear friends, King had no close human intimates when he first became prime minister in 1922. Pat became his confidant, listening to his problems and providing emotional support. When the Second World War was declared, King wrote in his secret diary that "little Pat always seems to me a sort of symbol of my mother," and that his mother "was giving me assurance of being at my side at this most critical of all the moments" (Coren 2002). Two years later, when Pat was 17 years old, he was deaf, growing blind, sick, and dying. In his last hours, King postponed a war-committee meeting so that he could hold Pat in his arms. He whispered messages of love to the dog, and asked him to be an emissary, taking them to his mother and other relatives and friends in heaven. Then he sang the hymn "God be with you till we meet again" as the little animal's body grew cold.

Chapter One: Evolutionary Matters

1. Here, an animal who is said to have "good genes" is one who has lived and reproduced successfully to a ripe old age. It must not be confused with the term as used, or rather misused, by evolutionary psychologists. For a discussion of this matter, see my book *"Love of Shopping" Is Not a Gene* (2005, 172–175).

2. Susan McElroy (2004) recalled asking one of the scientists, who oversaw the wolf transfer and cared passionately about the vicissitudes of Number 9, why the animals were given numbers rather than names. "This is an experimental group of animals. The survival of one is not as important as the survival of the group. We see them as a group," a male zoologist explained. McElroy did not believe this for a minute. When the scientists were not around, one of the men's wives told her that the men were attached to the animals, but they knew that some would be killed, and thought it would be less painful to lose an animal who did not have a name. McElroy continued to be upset, because she felt that the wolves were being slighted by remaining nameless. Later, she brightened up when she was reminded that in early times, "the word for God was a series of unpronounceable letters, because the true name of the Holy One is beyond words."

More prosaically, Elizabeth Marshall Thomas (1994) recalled that when she worked on an elephant project with Katy Payne in 1986, they also were ordered never to name the elephants they studied, but to give them numbers instead, which would make their work seem more scientific. "Wild animals have no names, it is true, but they certainly have no numbers either. With this in mind we named the elephants anyway, despite the warnings, because names are much easier to remember than numbers."

Chapter Two: Sociality, Media, and Variability

1. In a chapter related to media, it is of interest that magazines and television focus in general on young women and men, whereas when they consider non-human animals, they valorize older individuals. One example is the ancient tor-

toise, a friend of Charles Darwin, who died recently—to much fanfare—at about 175 years of age (Rook 2006). Another is the famous elephant Raja, celebrated for 58 years as Sri Lanka's "most venerated moving monument" (Alexander 2000). In Kandy, he led the annual summer parade of more than 50 extravagantly bejeweled elephants honoring Buddha. He carried a gold casket, in which resided the golden tooth relic of Buddha, on his back. When he died at age 65, Buddhist monks paid him tribute and thousands came to mourn him. He is now stuffed and mounted in a museum beside the Temple of the Tooth Relic.

Then there was Angel, an old cheetah who lived in the Cincinnati Zoo in the 1980s (Angel Fund 2006). For 12 years, she and Cathryn Hilker formed a team to provide education about cheetahs and their homeland—the Angel Fund website connected with over one million people during Angel's long life. Cheetah numbers have declined catastrophically in Africa (100,000 in 1900 and only about 11,000 today), so in 1992 the Angel Fund was set up in her memory. Its mission, through the Cheetah Conservation Fund, is to help preserve the species both in the wild and with breeding programs in captivity. One of its success stories has been providing farmers in Namibia, southwest Africa, with Anatolian shepherd guard dogs who protect livestock from cheetahs, so that the cheetahs do not have to face lethal retribution from farmers. Dogs saving cheetahs!

Chapter Three: The Wisdom of Elders

1. Recent research shows that learning, and experience gained throughout one's life, cause physical changes in the human brain. "The brain cells (called neurons) in the parts of the brain that an older person has used continuously would look like a dense forest of thickly branched trees, compared with the thinner and less dense forest of a young brain. This neural density is the physical basis for the skills of accomplished older adults" (Cohen 2005). The brains of elephants and other nonhuman animals presumably develop in a similar way during an animal's lifetime.

Chapter Four: Leaders

1. Eve was called Scar by some researchers (Rose 2000).

Chapter Five: Teaching and Learning

1. Bennett Galef (1992) notes that the terms "tradition" and "culture" are both used for nonhuman animals, implying that new discoveries and rites can be

carried down from one generation to the next. He argues persuasively that such culture is not homologous with human culture, which involves language and the arts, and should not be treated as such.

Chapter Eight: The Fall of Titans

1. I so dislike using this word, implying that a male can own females, that I wrote a book entitled *Harems and Other Horrors: Sexual Bias in Behavioral Biology* (1983). Although one male may often be seen with a group of females, close study usually shows that the males do not have mating rights if the females refuse them. A red deer stag in the rutting season does not gather compliant hinds around him, but each day acts rather like a sheep dog, trying to herd together as many females as possible, while at the same time keeping other stags at bay (Darling 1969). For the hamadryas baboons, however, the word "harem" is almost applicable. Males segregate their females from a young age, and keep a close watch on their activities.

2. Habituation can be defined as being present when a person can stand or sit within 10 m of an animal, without the animal looking at him or her more often than at troop members (Rasmussen 1991).

Chapter Twelve: Grandmothers

1. Peccei (2001) has written an extensive critique of the grandmother hypothesis, old and new versions.

Chapter Thirteen: Sexy Seniors

1. Some religious extremists insist that there is no homosexual behavior in nonhuman animals, thereby inferring that it is an evil practice in human beings, but they are wrong. Hundreds of animal species, both male and female, have same-sex sexual activity, as Bruce Bagemihl (1999) and I (1984) have documented in detail.

Chapter Fifteen: Adapting and Not Adapting

1. In general, older sea mammals tend to have more scars and gouges on their bodies from past encounters with boats than younger ones do. Ironically, these marks of past injury serve to identify these individuals, and can lead to further harassment by humans (Lord 2004).

Chapter Sixteen: All Passion Spent

1. Older primates may grow fat with greed if they are allowed to overeat. At the huge buffet of goodies set out each year in Lopburi, Thailand, for free-ranging long-tailed macaques—because it is said to bring good luck (and incidentally attracts tourists)—monkeys have wolfed down huge quantities of food over the years. Some individuals have become so obese that they are no longer able to move normally (Grice 2006).

REFERENCES

Abegglen, Jean-Jacques. 1985. *On Socialization in Hamadryas Baboons*. Cranbury, NJ: Associated University Presses.

Ackerman, Diane. 1998. The moon by whale light. In Linda Hogan, Deena Metzger, and Brenda Peterson, eds., *Intimate Nature: The Bond between Women and Animals*, 304–308. New York: Fawcett Books.

Alcock, John. 1988. *The Kookaburra's Song: Exploring Animal Behavior in Australia*. Tucson: University of Arizona Press.

Alexander, Shana. 2000. *The Astonishing Elephant*. New York: Random House.

Altmann, Jeanne. 1980. *Baboon Mothers and Infants*. Cambridge, MA: Harvard University Press.

Amoss, Pamela T. 1981. Coast Salish elders. In Pamela T. Amoss and Stevan Harrell, eds., *Other Ways of Growing Old*, 227–247. Stanford, CA: Stanford University Press.

Amoss, Pamela T. and Stevan Harrell, eds. 1981. *Other Ways of Growing Old: Anthropological Perspectives*. Stanford, CA: Stanford University Press.

Anderson, Connie M. 1986. Female age: Male preference and reproductive success in primates. *International Journal of Primatology* 7,3: 305–326.

Angel Fund. 2006. www.cincyzoo.org/conservation/GlobalConservation/cheetah/AngelFund/angelfundrev.html.

Apio, Ann, Martin Plath, Ralph Tiedemann, and Torsten Wronski. 2007. Age-dependent mating tactics in male bushbuck (*Tragelaphus scriptus*). *Behaviour* 144,5: 585–610.

Archie, Elizabeth A., Thomas A. Morrison, Charles A.H. Foley, Cynthia Moss, and Susan C. Alberts. 2006. Dominance rank relationships among wild female African elephants, *Loxodonta africana*. *Animal Behaviour* 71,1: 117–127.

Askins, Renée. 2002. *Shadow Mountain: A Memoir of Wolves, A Woman, and the Wild*. New York: Anchor Books.

Atsalis, Sylvia and Susan W. Margulis. 2006. Sexual and hormonal cycles in geriatric *Gorilla gorilla gorilla*. *International Journal of Primatology* 27,6: 1663–1687.

Atsalis, Sylvia, Susan W. Margulis, Astrid Bellem, and Nadja Wielebnowski.

2004. Sexual behavior and hormonal estrus cycles in captive aged lowland gorillas (*Gorilla gorilla*). *American Journal of Primatology* 62: 123–132.

Bagemihl, Bruce. 1999. *Biological Exuberance: Animal Homosexuality and Natural Diversity*. New York: St. Martin's Press.

Bailey, Theodore N. 1993. *The African Leopard: Ecology and Behavior of a Solitary Felid*. New York: Columbia University Press.

Balcombe, Jonathan. 2006. *The Pleasurable Kingdom: Animals and the Nature of Feeling Good*. London: Macmillan.

Barclay, Robert M.R. and Lawrence D. Harder. 2003. Life histories of bats: Life in the slow lane. In Thomas H. Kunz and M. Brock Fenton, eds., *Bat Ecology*, 209–253. Chicago: University of Chicago Press.

Bass, Rick. 1993. The way wolves are. In John A. Murray, ed., *Out Among the Wolves: Contemporary Writings on the Wolf*, 177–187. Vancouver: Whitecap Books.

Beehner, J.C., T.J. Bergman, D.L. Cheney, R.M. Seyfarth, and P.L. Whitten. 2005. The effect of new alpha males on female stress in free-ranging baboons. *Animal Behaviour* 69,5: 1211–1221.

Bekoff, Marc. 2005. E-mail re. a report by Janet Spittler on "Gorilla religiosus." March 3.

———. 2006. Would you do it to your dog? (e-mail report). September 28.

———. 2007. *Emotional Lives of Animals: A Leading Scientist Explores Animal Joy, Sorrow and Empathy—And Why They Matter*. Novato, CA: New World Library.

Berthold, Peter. 1996. *Control of Bird Migration*. London: Chapman and Hall.

Best, P.B., C.M. Schaeff, D. Reeb, and P.J. Palsboll. 2003. Composition and possible function of social groupings of southern right whales in South African waters. *Behaviour* 140,11–12: 1469–1494.

Biesele, Megan and Nancy Howell. 1981. "The old people give you life": Aging among !Kung hunter-gatherers. In Pamela T. Amoss and Stevan Harrell, eds., *Other Ways of Growing Old*, 77–98. Stanford, CA: Stanford University Press.

Birkhead, Tim R. and Anders Pape Møller. 1998. *Sperm Competition and Sexual Selection*. London: Academic Press.

Boguszewski, P. and J. Zagrodzka. 2002. Emotional changes related to age in rats—a behavioral analysis. *Behavior and Brain Research* 133,2: 323–332.

Borries, Carola. 1988. Patterns of grandmaternal behaviour in free-ranging Hanuman langurs (*Presbytis entellus*). *Human Evolution* 3,4: 239–260.

Bouwman, Karen M., Rene E. van Dijk, Jan J. Wijmenga, and Jan Komdeur. 2007. Older male reed buntings are more successful at gaining extrapair fertilizations. *Animal Behaviour* 73,1: 15–27.

Bradshaw, G.A., Allan N. Schore, Janine L. Brown, Joyce H. Poole, and Cynthia J. Moss. 2005. Concept elephant breakdown. *Nature* 433, 807.

Broussard, D.R., T.S. Risch, F.S. Dobson, and J.O. Murie. 2003. Senescence and age-related reproduction of female Columbian ground squirrels. *Journal of Animal Ecology* 72: 212–219.

Brown, Charles R. 1998. *Swallow Summer*. Lincoln: University of Nebraska Press.

Brownlee, Shannon, M. and Kenneth S. Norris. 1994. The acoustic domain. In Kenneth S. Norris, Bernd Würsig, Randall S. Wells, and Melany Würsig, eds., *The Hawaiian Spinner Dolphin*, 161–185. Berkeley: University of California Press.

Burger, Joanna. 2001. *The Parrot Who Owns Me: The Story of a Relationship*. New York: Random House.

Buss, Irven O. 1990. *Elephant Life: Fifteen Years of High Population Density*. Ames: Iowa State University Press.

Byers, John A. 2003. *Built for Speed: A Year in the Life of Pronghorn*. Cambridge, MA: Harvard University Press.

Cameron, Elissa Z., Wayne L. Linklater, Kevin J. Stafford, and Edward O. Minot. 2000. Aging and improving reproductive success in horses: Declining residual reproductive value or just older and wiser? *Behavioral Ecology and Sociobiology* 47: 243–249.

Campagna, Claudio, Claudio Bisioli, Flavio Quintana, Fabian Perez, and Alejandro Vila. 1992. Group breeding in sea lions: Pups survive better in colonies. *Animal Behaviour* 43: 541–548.

Carey, James R. 2003. *Longevity: The Biology and Demography of Life Span*. Princeton, NJ: Princeton University Press.

Caro, T.M. and M.D. Hauser. 1992. Is there teaching in nonhuman animals? *Quarterly Review of Biology* 67,2: 151–174.

Carr, Rosamond Halsey, with Ann Howard Halsey. 1999. *Land of a Thousand Hills: My Life in Rwanda*. New York: Viking.

Carrighar, Sally. 1965. *Wild Heritage*. Boston: Houghton Mifflin.

Chadwick, Douglas H. 1992. *The Fate of the Elephant*. San Francisco: Sierra Club Books.

Cheney, Dorothy L. and Robert M. Seyfarth. 1990. *How Monkeys See the World: Inside the Mind of Another Species*. Chicago: University of Chicago Press.

Clapham, Phil. 2004. *Right Whales: Natural History and Conservation*. Stillwater, MN: Voyageur Press.

Clotfelter, Ethan D., Alison M. Bell, and Kate R. Levering. 2004. The role of animal behaviour in the study of endocrine-disrupting chemicals. *Animal Behaviour* 68,4: 665–676.

Clutton-Brock, T.H. 1984. Reproductive effort and terminal investment in itero-parous animals. *American Naturalist* 123,2: 212–229.

Clutter-Brock, T.H., F.E. Guinness, and S.D. Albon. 1982. *Red Deer Behaviour and Ecology of Two Sexes.* Edinburgh: Edinburgh University Press.

Cohen, Gene D. 2005. *The Mature Mind: The Positive Power of the Aging Brain.* New York: Basic Books.

Connor, Richard C. and Dawn Micklethwaite Peterson. 1994. *The Lives of Whales and Dolphins.* New York: Henry Holt.

Coren, Stanley. 1994. *The Intelligence of Dogs: Canine Consciousness and Capabilities.* New York: Free Press.

————. 2002. *The Pawprints of History.* New York: Free Press.

Cote, Steve D. and Marco Festa-Blanchet. 2001. Reproductive success in female mountain goats: The influence of age and social rank. *Animal Behaviour* 62,1: 173–181.

Couturier, Lisa. 2005. *The Hopes of Snakes and Other Tales from the Urban Landscape.* Boston: Beacon Press.

Crandall, Lee S. 1966. *A Zoo Man's Notebook.* Chicago: University of Chicago Press.

Creel, Scott. 2001. Social dominance and stress hormones. *Trends in Ecology and Evolution* 16,9: 491–497.

————. 2005. Dominance, aggression, and glucocorticoid levels in social carnivores. *Journal of Mammalogy* 86,2: 255–264.

Dagg, Anne Innis. 1970. Tactile encounters in a herd of captive giraffe. *Journal of Mammalogy* 51: 279–287.

————. 1983. *Harems and Other Horrors: Sexual Bias in Behavioral Biology.* Waterloo, ON: Otter Press.

————. 1984. Homosexual behaviour and female-male mounting in mammals— a first survey. *Mammal Review* 14: 155–185.

————. 2005. *"Love of Shopping" Is Not a Gene: Problems with Darwinian Psychology.* Montreal: Black Rose Press.

Dagg, Anne Innis and J. Bristol Foster. 1976. *The Giraffe: Its Biology, Behavior, and Ecology.* New York: Van Nostrand Reinhold.

Dahlberg, Carrie Peyton. 2007. Old monkeys, new memory clues: Research on primate aging could aid humans. *Sacramento Bee,* January 22, A1. http://www.sacbee.com/303/story/83452.html.

Darling, F. Fraser. 1969. *A Herd of Red Deer: A Study in Animal Behaviour.* London: Oxford University Press. (Orig. pub. 1937.)

Davidson, Terry. 2005. A more personal way to say goodbye. *National Post* (Toronto). December 31, TO22.

Delean, Paul. 2007. Unbridled love for a racehorse. *Montreal Gazette.* July 3.

DelGiudice, Glenn D., Mark S. Lenarz, and Michelle Carstensen Powell. 2007. Age-specific fertility and fecundity in northern free-ranging white-tailed deer: Evidence for reproductive senescence? *Journal of Mammalogy* 88,2: 427–435.

DeRousseau, C. Jean, Laszlo Z. Bito, and Paul L. Kaufman. 1986. Age-dependent impairments of the rhesus monkey visual and musculoskeletal systems and apparent behavioral consequences. In Richard G. Rawlins and Matt J. Kessler, eds., *The Cayo Santiago Macaques: History, Behavior and Biology*, 233–251. Albany: State University of New York Press.

Diamond, Jared. 1987. Learned specializations of birds. *Nature* 330,5: 16–17.

———. 2001. Unwritten knowledge. *Nature* 410: 521.

Discovery Channel. 2006. 176-year-old tortoise named Harriet passes on. June 26. http://dsc.discovery.com/news/2006/06/26/tortoise_ani.html?category=animals &guid=20060626114500.

Doidge, Norman. 2007. *The Brain that Changes Itself.* New York: Viking.

Dolhinow, Phyllis, James J. McKenna, and Julia Vonder Haar Laws. 1979. Rank and reproduction among female langur monkeys: Aging and improvement (they're not just getting older, they're getting better). *Aggressive Behaviour* 5: 19–30.

Douglas-Hamilton, Iain and Oria Douglas-Hamilton. 1975. *Among the Elephants.* London: Collins and Harvill Press.

Dutcher, Jim and Jamie Dutcher. 2002. *Wolves at Our Door.* New York: Pocket Books.

Engel, Cindy. 2002. *Wild Health: How Animals Keep Themselves Well and What We Can Learn from Them.* Boston: Houghton Mifflin.

Engh, Anne L., Jacinta C. Beehner, Thore J. Bergman, Patricia L. Whitten, Rebekah R. Hoffmeier, Robert M. Seyfarth, and Dorothy L. Cheney. 2006. Behavioural and hormonal responses to predation in female chacma baboons (*Papio hamadryas ursinus*). *Proceedings of the Royal Society B: Biological Sciences* 273: 707–712.

Ericsson, Göran, Kiell Wallin, John P. Ball, and Martin Broberg. 2001. Age-related reproductive effort and senescence in free-ranging moose, *Alces alces*. *Ecology* 82,6: 1613–1620.

Fairbanks, L.A. and M.T. McGuire. 1986. Age, reproductive value, and dominance-related behaviour in vervet monkey females: Cross-generational influences on social relationships and reproduction. *Animal Behaviour* 34: 1710–1721.

Fedigan, Linda Marie. 1991. History of the Arashiyama West Japanese macaques in Texas. In Linda Marie Fedigan and Pamela J. Asquith, eds., *The Monkeys of Arashiyama*, 54–73. Albany: State University of New York Press.

Fedigan, Linda Marie and Mary S. McDonald Pavelka. 2001. Is there adaptive value to reproductive termination in Japanese macaques? A test of maternal investment hypotheses. *International Journal of Primatology* 22,2: 109–125.

Findley, Timothy. 1990. *Inside Memory.* Toronto: HarperCollins.

Fleetwood, Mary Anne. 1998. An interview with Linda Tellington-Jones. In Linda Hogan, Deena Metzger, and Brenda Peterson, eds., *Intimate Nature: The Bond between Women and Animals*, 203–208. New York: Fawcett Books.

Fossey, Dian. 1983. *Gorillas in the Mist.* Boston: Houghton Mifflin.

Fox, Michael W. 1980. *The Soul of the Wolf.* Boston: Little, Brown.

Fukuyama, Francis. 1988. Women and the evolution of world politics. *Foreign Affairs* (September/October): 24–40.

Galef, Bennett G. 1992. The question of animal culture. *Human Nature* 3,2: 157–178.

Galef, Bennett G., Elaine E. Whiskin, and Gwen Dewar. 2005. A new way to study teaching in animals: Despite demonstrable benefits, rat dams do not teach their young what to eat. *Animal Behaviour* 70,1: 91–96.

Gauthier-Pilters, Hilde and Anne Innis Dagg. 1981. *The Camel: Its Ecology, Behavior and Relationship with Man.* Chicago: University of Chicago Press.

Geist, Valerius. 1971. *Mountain Sheep: A Study in Behavior and Evolution.* Chicago: University of Chicago Press.

Glavin, Terry. 2006. According to science, whales are for the killing. *Globe and Mail* (Toronto), July 1, F7.

Goodall, Jane. 1986. *The Chimpanzees of Gombe: Patterns of Behavior.* Cambridge, MA: Belknap Press.

Gordon, Jonathan. 1998. *Sperm Whales.* Stillwater, MN: Voyageur Press.

Gorman, Anna. 2006. Death of Gita renews calls to move elephants to sanctuary: Activists demonstrate at the L.A. Zoo to protest its plans to keep two remaining pachyderms. *Los Angeles Times,* June 12, B3.

Grandin, Temple and Catherine Johnson. 2005. *Animals in Translation: Using the Mysteries of Autism to Decode Animal Behavior.* New York: Scribner.

Grice, Samantha. 2006. Curious gorge. *National Post* (Toronto), November 28, AL8.

Grogan, John. 2005. *Grogan and Me: Life and Love with the World's Worst Dog.* New York: William Morrow.

Gröning, Karl and Martin Saller. 1999. *Elephants: A Cultural and Natural History.* Cambridge: First Edition Translations, from the German.

Gunderson, Harvey L. 1976. *Mammalogy.* New York: McGraw-Hill.

Gunji, Harumoto, Kazuhiko Hosaka, Michael A. Huffman, Kenji Kawanaka, Akiko Matsumoto-Oda, Yuzuru Hamada, and Toshisada Nishida. 2003. Extraordinarily low bone mineral density in an old female chimpanzee (*Pan tro-*

glodytes schweinfurthii) from the Mahale Mountains National Park. *Primates* 44: 145–149.

Haas, Emmy. 1967. *Pride's Progress: The Story of a Family of Lions.* New York: Harper and Row.

Hanby, Jeannette. 1982. *Lions Share.* Boston: Houghton Mifflin.

Hatkoff, Craig, Isabella Hatkoff, and Paula Kahumbu. 2005. *Owen and Mzee: The True Story of a Remarkable Friendship.* New York: Scholastic Press.

Hauser, Marc D. 1988. Invention and social transmission: New data from wild vervet monkeys. In R.W. Byrne and A. Whiten, eds., *Machiavellian Intelligence: Social Expertise and the Evolution of Intellect in Monkeys, Apes, and Humans,* 327–343. Oxford: Oxford University Press.

Hauser, Marc D. and Gary Tyrrell. 1984. Old age and its behavioral manifestations: A study on two species of macaque. *Folia Primatologica* 43: 24–35.

Havelka, M.A. and J.S. Millar. 2004. Maternal age drives seasonal variation in litter size of *Peromyscus leucopus. Journal of Mammalogy* 85,5: 940–947.

Hawkes, Kristen. 2003. Grandmothers and the evolution of human longevity. *American Journal of Human Biology* 15: 380–400.

Hawkes, Kristen, J.F. O'Connell, and N.G. Blurton-Jones. 1997. Hadza women's time allocation, offspring provisioning and the evolution of long postmenopausal life spans. *Current Anthropology* 38: 551–577.

Hediger, Heini. 1968. *The Psychology and Behaviour of Animals in Zoos and Circuses.* Toronto: General Publishing.

Herman, L. 1980. Cognitive characteristics of dolphins. In L. Herman, ed., *Cetacean Behavior: Mechanisms and Functions,* 363–430. New York: Wiley-Interscience.

Hinde, Gerald. 1992. *Leopard.* London: HarperCollins.

Hinton, H.E. and A.M.S. Dunn. 1967. *Mongooses: Their Natural History and Behaviour.* Edinburgh: Oliver and Boyd.

Hoogland, John L. 1995. *The Black-Tailed Prairie Dog: Social Life of a Burrowing Mammal.* Chicago: University of Chicago Press.

Howard, Carol J. 1995. *Dolphin Chronicles.* New York: Bantam Books.

Hoyt, Erich. 1990. *Orca: A Whale Called Killer.* Toronto: Camden House.

Hrdy, Sarah Blaffer. 1977. *The Langurs of Abu: Female and Male Strategies of Reproduction.* Cambridge, MA: Harvard University Press.

———. 1981. "Nepotists" and "altruists": The behavior of old females among macaques and langur monkeys. In Pamela T. Amoss and Stevan Harrell, eds., *Other Ways of Growing Old,* 59–76. Stanford, CA: Stanford University Press.

———. 1999. *Mother Nature: A History of Mothers, Infants, and Natural Selection.* New York: Pantheon Books.

Huffman, Michael A. 1990. Some socio-behavioral manifestations of old age. In Toshisada Nishida, ed., *The Chimpanzees of the Mahale Mountains: Sexual and Life History Strategies,* 237–255. Tokyo: University of Tokyo Press.

Hyman, J., M. Hughes, W.A. Searcy, and S. Nowicki. 2004. Individual variation in the strength of territory defense in male song sparrows: Correlates of age, territory tenure, and neighbor aggressiveness. *Behaviour* 141,1: 15–27.

Innis, Harold Adams. 1952. Autobiography, 1894 to 1922, typescript, p. 14.

Jackson, Peter. 2007. The elephants' farewell. *Sunday Times* (London), January 28, magazine section.

Jensen, Gordon D., F.L. Blanton, and David H. Gribble. 1980. Older monkeys' (*Macaca radiata*) response to new group formation: Behavior, reproduction and mortality. *Experimental Gerontology* 15: 399–406.

Johnson, Christine M. and Kenneth S. Norris. 1994. Social behavior. Kenneth S. Norris, Bernd Würsig, Randall S. Wells, and Melany Würsig, eds., *The Hawaiian Spinner Dolphin,* 243–286. Berkeley: University of California Press.

Johnstone-Scott, Richard. 1995. *Jambo: A Gorilla's Story.* London: Michael O'Mara Books.

Jolly, Alison. 2004. *Lords and Lemurs: Mad Scientists, Kings with Spears, and the Survival of Diversity in Madagascar.* Boston: Houghton Mifflin.

Kano, Takayoshi. 1992. *The Last Ape: Pygmy Chimpanzee Behavior and Ecology.* Stanford, CA: Stanford University Press.

Kaplan, Gisela and Lesley J. Rogers. 2000. *The Orangutans.* Cambridge, MA: Perseus Publishing.

Kawai, Masao. 1965. Newly-acquired pre-cultural behavior of the natural troop of Japanese monkeys on Koshima Islet. *Primates* 6: 1–30.

Kawanaka, Kenji. 1990. Alpha males' interactions and social skills. In Toshisada Nishida, ed., *The Chimpanzees of the Mahale Mountains: Sexual and Life History Strategies,* 171–187. Tokyo: University of Tokyo Press.

Kempermann, Gerd, Daniela Gast, and Fred H. Gage. 2002. Neuroplasticity in old age: Sustained fivefold induction of hippocampal neurogenesis by long-term environmental enrichment. *Annals of Neurology* 52: 135–143.

Kim, Sin-Yeon, Roxana Torres, Cristina Rodriguez, and Hugh Drummond. 2007. Effects of breeding success, mate fidelity and senescence on breeding dispersal of male and female blue-footed boobies. *Journal of Animal Ecology* 76: 471–479.

Kipps, Clare. 1953. *Sold for a Farthing.* London: Frederick Muller.

Knudtson, Peter. 1996. *Orca: Visions of the Killer Whale.* Vancouver: Greystone Books.

Kolata, Gina. 2007. These runners are stronger, faster and older. *National Post* (Toronto), September 15, WP9.

Kumar, Palash. 2006. Lions dying in Indian zoo after failed experiment. Reuters. September 17. http://bigcatnews.blogspot.com/2006_09_01_archive.html and http://www.asiatic-lion.org/news/news-0619.html.

Kummer, Hans. 1995. *In Quest of the Sacred Baboon: A Scientist's Journey.* Princeton, NJ: Princeton University Press.

Lahdenperä, Mirkka, Virpi Lummaa, Samuli Helle, Marc Tremblay, and Andrew F. Russell. 2004. Fitness benefits of prolonged post-reproductive lifespan in women. *Nature* 428, 178–181.

Larousse Encyclopedia of Animal Life. 1967. London: Hamlyn.

Lawick-Goodall, Hugo van and Jane van Lawick-Goodall. 1970. *Innocent Killers.* London: Collins.

Laws, R.M., I.S.C. Parker, and R.C.B. Johnstone. 1975. *Elephants and Their Habitats: The Ecology of Elephants in North Bunyoro, Uganda.* Oxford: Clarendon Press.

Lehmann, Julia, Gisela Fickenscher, and Christophe Boesch. 2006. Kin biased investment in wild chimpanzees. *Behaviour* 143,8: 931–955.

Ligon, J.D. and S.H. Ligon. 1990. Great woodhoopoes: Life history traits and sociality. In Peter B. Stacey and Walter D. Koenig, eds., *Cooperative Breeding in Birds,* 31–65. Cambridge: Cambridge University Press.

Lilly, John Cunningham. 1975. *Lilly on Dolphins: Humans of the Sea.* New York: Anchor Books.

Lindau, Stacy Tessler, L. Philip Schumm, Edward O. Laumann, Wendy Levinson, Colm A. O'Muircheartaigh, and Linda Waite. 2007. A study of sexuality and health among older adults in the United States. *New England Journal of Medicine* 357,8: 762–774.

Linden, Eugene. 1999. *The Parrot's Lament and Other True Tales of Animal Intrigue, Intelligence, and Ingenuity.* New York: Penguin.

Logan, Kenneth A. and Linda L. Sweanor. 2001. *Desert Puma: Evolutionary Ecology and Conservation of an Enduring Carnivore.* Washington, DC: Island Press.

Lord, Nancy. 2004. *Beluga Days: Tracking a White Whale's Truth.* New York: Counterpoint.

Macdonald, David. 2000. Night school. In Marc Bekoff, ed., *The Smile of a Dolphin,* 46–49. New York: Discovery Books.

Maples, E.G., M.M. Haraway, and C.W. Hutto. 1989. Development of coordinated singing in a newly formed siamang pair (*Hylobates syndactylus*). *Zoo Biology* 8,4: 367–378.

Marsh, Helene and Toshio Kasuya. 1991. An overview of the changes in the role of a female pilot whale with age. In Karen Pryor and Kenneth S. Norris, eds., *Dolphin Societies: Discoveries and Puzzles,* 281–285. Berkeley: University of California Press.

Marzluff, J.M. and R.P. Balda. 1990. Pinyon jays: Making the best of a bad situation by helping. In Peter B. Stacey and Walter D. Koenig, eds., *Cooperative Breeding in Birds*, 197–237. Cambridge: Cambridge University Press.

Masson, Jeffrey Moussaieff. 1997. *Dogs Never Lie about Love: Reflections on the Emotional World of Dogs*. New York: Crown Publishers.

———. 2006. *Altruistic Armadillos, Zenlike Zebras*. New York: Random House.

Masson, Jeffrey Moussaieff and Susan McCarthy. 1995. *When Elephants Weep: The Emotional Lives of Animals*. New York: Delacorte Press.

Masters, Brian. 1988. *The Passion of John Aspinall*. London: Jonathan Cape.

McCarthy, Susan. 2004. *Becoming Tiger: How Baby Animals Learn to Live in the Wild*. New York: HarperCollins.

McComb, Karen, Cynthia Moss, Sarah M. Durant, Lucy Baker, and Soila Sayialel. 2001. Matriarchs as repositories of social knowledge in African elephants. *Science* 292,5516: 491–494.

McComb, Karen, David Reby, Lucy Baker, Cynthia Moss, and Soila Sayialel. 2003. Long-distance communication of acoustic cues to social identity in African elephants. *Animal Behaviour* 65,2: 317–329.

McCracken, Harold. 2003. *The Beast That Walks Like Man: The Story of the Grizzly Bear*. Lanham, MD: Roberts Rinehart. (Orig. pub. 1955.)

McElroy, Susan Chernak. 1996. *Animals as Teachers and Healers*. New York: Ballantine Books.

———. 2004. *All My Relations: Living with Animals as Teachers and Healers*. Novato, CA: New World Library.

McIntyre, Rick. 1993. The East Fork pack. In John A. Murray, ed., *Out Among the Wolves: Contemporary Writings on the Wolf*, 189–192. Vancouver: Whitecap Books.

Mech, L. David. 1966. *The Wolves of Isle Royale*. Washington, DC: U.S. Government Printing Office.

Meltzoff, A.N. 1988. Imitation, objects, tools, and the rudiments of language in human ontogeny. *Human Evolution* 3,1–2: 45–64.

Melville, Herman. 2001. *Moby-Dick*. 2nd ed. New York: Norton. (Orig. pub. 1851.)

Meredith, Martin. 2007. Like humans, like elephants. *Conservation Magazine* 8,1: 1–2.

Mitani, M. 1986. Voiceprint identification and its application to sociological studies of wild Japanese monkeys (*Macaca fuscata*). *Primates* 27: 397–412.

Mizuhara, Hiroki. 1964. Social changes of Japanese monkey troops in the Takasakiyama. *Primates* 5: 29–51.

Morton, Alexandra. 2002. *Listening to Whales: What the Orcas Have Taught Us*. New York: Ballantine Books.

Moss, Cynthia. 1975. *Portraits in the Wild: Behavior Studies of East African Mammals*. Boston: Houghton Mifflin.

———. 1988. *Elephant Memories: Thirteen Years in the Life of an Elephant Family*. New York: William Morrow.

———. 1998. Elephant memories. In Linda Hogan, Deena Metzger, and Brenda Peterson, eds., *Intimate Nature: The Bond between Women and Animals*, 115–125. New York: Fawcett Books.

Mowat, Farley. 1988. *Virunga: The Passion of Dian Fossey*. Toronto: McClelland and Stewart. (Orig. pub. 1987.)

Muller, Martin N., Melissa Emery Thompson, and Richard W. Wrangham. 2006. Male chimpanzees prefer mating with old females. *Current Biology* 16,22: 2234–2238.

Mysterud, Atle, Erling J. Solberg, and Nigel G. Yoccoz. 2005. Ageing and reproductive effort in male moose under variable levels of intrasexual competition. *Journal of Animal Ecology* 74: 742–754.

Nakamichi, Masayuki. 1984. Behavioral characteristics of old female Japanese monkeys in a free-ranging group. *Primates* 25,2: 192–203.

———. 1991. Behavior of old females: Comparisons of Japanese monkeys in the Arashiyama East and West groups. In Linda Marie Fedigan and Pamela J. Asquith, eds., *The Monkeys of Arashiyama*, 175–193. Albany: State University of New York Press.

———. 2003. Age-related differences in social grooming among adult female Japanese monkeys (*Macaca fuscata*). *Primates* 44: 239–246.

Nakamichi, Masayuki, April Silldorff, Crystal Bringham, and Peggy Sexton. 2004. Baby transfer and other interactions between its mother and grandmother in a captive social group of lowland gorillas. *Primates* 45: 73–77.

Nakamichi, Masayuki and K. Yamada. 2007. Long-term grooming partnerships between unrelated adult females in a free-ranging group of Japanese monkeys (*Macaca fuscata*). *American Journal of Primatology* 69: 652–663.

Nason, James D. 1981. Respected elder or old person: Aging in a Micronesian Community. In Pamela T. Amoss and Stevan Harrell, eds., *Other Ways of Growing Old*, 155–173. Stanford, CA: Stanford University Press.

National Post (Toronto). 2006. Patsy the African elephant (1966–2006). July 26, A7.

Nishida, Toshisada, Hiroyuki Takasaki, and Yukio Takahata. 1990. Demography and reproductive profiles. In Toshisada Nishida, ed., *The Chimpanzees of the Mahale Mountains: Sexual and Life History Strategies*, 63–97. Tokyo: University of Tokyo Press.

Nishida, Toshisada and sixteen others. 2003. Demography, female life history, and reproductive profiles among the chimpanzees of Mahale. *American Journal of Primatology* 59: 99–121.

Nollman, Jim. 1999. *The Charged Border Where Whales and Humans Meet.* New York: Henry Holt.

Norris, Kenneth S. 1974. *The Porpoise Watcher.* New York: Norton.

———. 1991. *Dolphin Days: The Life and Times of the Spinner Dolphin.* New York: Norton.

Norris, Kenneth S. and Karen Pryor. 1991. Some thoughts on grandmothers. In Karen Pryor and Kenneth S. Norris, eds., *Dolphin Societies: Discoveries and Puzzles,* 287–289. Berkeley: University of California Press.

Norris, Kenneth S., Bernd Würsig, and Randall S. Wells. 1994. Aerial behavior. In Kenneth S. Norris, Bernd Würsig, Randall S. Wells, and Melany Würsig, eds., *The Hawaiian Spinner Dolphin,* 103–121. Berkeley: University of California Press.

North, Sterling. 1966. *Raccoons Are the Brightest People.* New York: Dutton.

O'Connell-Rodwell, Caitlin E., Jason D. Wood, Roland Gunther, Simon Klemperer, Timothy C. Rodwell, Sunil Puria, Robert Sapolsky, Colleen Kinzley, Byron T. Arnason, and Lynette A. Hart. 2004. Elephant low-frequency vocalizations propagate in the ground and seismic playbacks of these vocalizations are detectable by wild African elephants (*Loxodonta africana*). *Journal of the Acoustical Society of America* 115,5: 2554.

Olson, Sigurd. 1987. *Songs of the North.* New York: Penguin.

Orell, Markku and Eduardo J. Belda. 2002. Delayed cost of reproduction and senescence in the willow tit *Parus montanus. Journal of Animal Ecology* 71: 55–64.

Owens, Anne Marie. 2002. World's oldest known bird found—still looking for sex. *National Post* (Toronto), April 20, A2.

Owens, Mark and Delia Owens. 2006. *Secrets of the Savanna: Twenty-Three Years in the African Wilderness Unraveling the Mysteries of Elephants and People.* Boston: Houghton Mifflin.

Packer, Craig, Marc Tatar, and Anthony Collins. 1998. Reproductive cessation in female mammals. *Nature* 392: 807–811.

Pärt, Tomas, Lars Gustafsson, and Juan Moreno. 1992. "Terminal investment" and a sexual conflict in the collared flycatcher (*Ficedula albicollis*). *American Naturalist* 140,5: 868–882.

Paul, Andreas, Jutta Kuester, and Doris Podzuweit. 1993. Reproductive senescence and terminal investment in female Barbary macaques (*Macaca sylvanus*) at Salem. *International Journal of Primatology* 14,1: 105–124.

Pavelka, Mary S. McDonald. 1990. Do old female monkeys have a specific social role? *Primates* 31,3: 363–373.

Pavelka, Mary S. McDonald, Linda M. Fedigan, and Sandra Zohar. 2002. Availability and adaptive value of reproductive and postreproductive Japanese macaque mothers and grandmothers. *Animal Behaviour* 64,3: 407–414.

Payne, Katy. 1998. *Silent Thunder in the Presence of Elephants*. New York: Simon and Schuster.

Payne, Roger. 1995. *Among Whales*. New York: Scribner.

Pearce, Tralee. 2007. $12 million dog. *Globe and Mail* (Toronto), August 30, L1, L5.

Peccei, Jocelyn Scott. 2001. A critique of the grandmother hypotheses: Old and new. *American Journal of Human Biology* 13: 434–452.

Pennisi, Elizabeth. 2001. Elephant matriarchs tell friend from foe. *Science* 292,5516: 417–418. http://www.sciencemag.org/cgi/content/summary/292/5516/417.

Petersen, David. 1995. *Ghost Grizzlies: Does the Great Bear Still Haunt Colorado?* New York: Henry Holt.

Peterson, Brenda. 1998. Apprenticeship to animal play. In Linda Hogan, Deena Metzger, and Brenda Peterson, eds., *Intimate Nature: The Bond between Women and Animals*, 428–437. New York: Fawcett Books.

———. 2001. *Build Me an Ark*. New York: Norton.

Podulka, Sandy, Ronald W. Rohrbaugh, and Rick Bonney. 2004. *Handbook of Bird Biology*. 2nd ed. Ithaca, NY: Cornell University Press.

Poesel, Angelika, P. Kunc Hansjoerg, Katharina Foerster, Arild Johnsen, and Bart Kempenaers. 2006. Early birds are sexy: Male age, dawn song and extrapair paternity in blue tits, *Cyanistes* (formerly *Parus*) *caeruleus*. *Animal Behaviour* 72,3: 531–538.

Poole, Alan. 1989. *Ospreys: A Natural and Unnatural History*. Cambridge: Cambridge University Press.

Poole, Joyce. 1996. *Coming of Age with Elephants*. London: Hodder and Stoughton.

Presty, Sharon K., Jocelyne Bachevalier, Lary C. Walker, Robert G. Struble, Donald L. Price, Mortimer Mishkin, and Linda C. Cork. 1987. Age differences in recognition memory of the rhesus monkey (*Macaca mulatta*). *Neurobiology of Aging* 8,5: 435–440.

Pugesek, Bruce H. 1981. Increased reproductive effort with age in the California gull (*Larus californicus*). *Science* 212,4496: 822–824.

Punzo, Fred and Sonia Chavez. 2003. Effect of aging on spatial learning and running speed in the shrew (*Cryptotis parva*). *Journal of Mammalogy* 84,3: 1112–1120.

Rasmussen, D.R. 1991. Observer influence on range use of *Macaca arctoides* after 14 years of observation? *Laboratory Primate Newsletter* 30,3: 6–11.

Record (Kitchener, ON). 1999. Seventy-two-year-old elephant dies of grief for her friend. May 6, A8.

Record (Kitchener, ON). 2006. Zoo euthanizes 46-year-old elephant. September 2, A12.

Record (Kitchener, ON). 2007. Zoo opens to allow goodbye to dying tiger. January 12, A4.

Redekop, Bill. 2006. Patches, the solitary pooch. *National Post* (Toronto), January 10, A2.

Reid, J.M., E.M. Bignal, S. Bignal, D.I. McCracken, and P. Monaghan. 2003. Age-specific reproductive performance in red-billed choughs *Pyrrhocorax pyrrhocorax:* Patterns and processes in a natural population. *Journal of Animal Ecology* 72: 765–776.

Ridley, Jo, Douglas W. Yu, and William J. Sutherland. 2005. Why long-lived species are more likely to be social: The role of local dominance. *Behavioral Ecology* 16,2: 358–363.

Roach, John. 2003. Biologists study evolution of animal cooperation. *National Geographic News,* July 9, 1–4. http://news.nationalgeographic.com/news/2003/07/0709_030709_socialanimals.html.

Robbins, Andrew M., Martha M. Robbins, Netzin Gerald-Steklis, and H. Dieter Steklis. 2006. Age-related patterns of reproductive success among female mountain gorillas. *American Journal of Physical Anthropology* 131: 511–521.

Robbins, Martha M., Andrew M. Robbins, Netzin Gerald-Steklis, and H. Dieter Steklis. 2005. Long-term dominance relationships in female mountain gorillas: Strength, stability and determinants of rank. *Behaviour* 142,6: 779–809.

Robertson, Raleigh J. and Wallace B. Rendell. 2001. A long-term study of reproductive performance in tree swallows: The influence of age and senescence on output. *Journal of Animal Ecology* 70: 1014–1031.

Roof, Katherine A., William D. Hopkins, M. Kay Izard, Michelle Hook, and Steven J. Schapiro. 2005. Maternal age, parity, and reproductive outcome in captive chimpanzees (*Pan troglodytes*). *American Journal of Primatology* 67: 199–207.

Rook, Katie. 2006. Harriet: 1830–2006. *National Post* (Toronto), June 24, A2.

Rose, Naomi A. 2000. A death in the family. In Marc Bekoff, ed., *The Smile of a Dolphin,* 144–145. New York: Discovery Books.

Ross, Mark C. 2001. *Dangerous Beauty.* New York: Hyperion.

Roth, George S., Julie A. Mattison, Mary Ann Ottinger, Mark E. Chachich, Mark A. Lane, and Donald K. Ingram. 2004. Aging in rhesus monkeys: Relevance to human health interventions. *Science* 305,5689: 1423–1426.

Rothschild, Bruce M. and Frank J. Ruhli. 2005. Comparison of arthritis characteristics in lowland *Gorilla gorilla* and mountain *Gorilla beringei*. *American Journal of Primatology* 66: 205–218.

Rowley, I. and E. Russell. 1990. Splendid Fairy-wrens: Demonstrating the importance of longevity. In Peter B. Stacey and Walter D. Koenig, eds., *Cooperative Breeding in Birds*, 1–30. Cambridge: Cambridge University Press.

Russell, Andy. 1977. *Andy Russell's Adventures with Wild Animals*. Edmonton: Best Printing.

Russell, Dick. 2001. *Eye of the Whale: Epic Passage from Baja to Siberia*. New York: Simon and Schuster.

Saino, Nicola, Roberto Ambrosini, Roberta Martinelli, and Anders Pape Møller. 2002. Mate fidelity, senescence in breeding performance and reproductive trade-offs in the barn swallow. *Journal of Animal Ecology* 71: 309–319.

Sapolsky, Robert M. 1990. Stress in the wild. *Scientific American* 1,262: 116–123.

— ——, 1994. *Why Zebras Don't Get Ulcers: A Guide to Stress, Stress-Related Diseases, and Coping* New York: Freeman and Co.

———. 1996. Why should an aged male baboon ever transfer troops? *American Journal of Primatology* 39: 149–157.

———. 1997. *The Trouble with Testosterone and Other Essays on the Biology of the Human Predicament*. New York: Simon and Schuster.

———. 2001. *A Primate's Memoir: A Neuroscientist's Unconventional Life among the Baboons*. Waterville, ME: Thorndike Press.

Savage, Candace. 1995. *Bird Brain: The Intelligence of Crows, Ravens, Magpies and Jays*. Vancouver: Greystone Books.

Schaller, George B. 1972a. *The Serengeti Lion*. Chicago: University of Chicago Press.

———. 1972b. *Serengeti: A Kingdom of Predators*. New York: Knopf.

Schmidt, Michael J. 1992. The elephant beneath the mask. In Jcheskel Shoshani, ed., *Elephants: Majestic Creatures of the Wild*, 92–97. Emmaus, PA: Rodale Press.

Schorger, A.W. 1955. *The Passenger Pigeon: Its Natural History and Extinction*. Norman: University of Oklahoma Press.

Scott, J.P. 1945. Social behavior, organization and leadership in a small flock of domestic sheep. *Comparative Psychology Monographs* 18,4: 1–29.

Segal, Nancy L. 1999. *Entwined Lives: Twins and What They Tell Us about Human Behavior*. New York: Dutton.

Shahrani, M. Nazif. 1981. Growing in respect: Aging among the Kirghiz of Afghanistan. In Paula T. Amoss and Stevan Harrell, eds., *Other Ways of Growing Old*, 175–191. Stanford, CA: Stanford University Press.

Sharp, Henry S. 1981. Old age among the Chipewyan. In Pamela T. Amoss and

Stevan Harrell, eds., *Other Ways of Growing Old*, 99–109. Stanford, CA: Stanford University Press.

Shortt, Terry. 1975. *Not as the Crow Flies*. Toronto: McClelland and Stewart.

Siebert, Charles. 2006. An elephant crackup? *New York Times*, October 8, magazine section.

Sikes, Sylvia K. 1971. *The Natural History of the African Elephant*. New York: American Elsevier Publishing.

Small, Meredith F. 1984. Aging and reproductive success in female *Macaca mulatta*. In Meredith F. Small, ed., *Female Primates: Studies by Women Primatologists*, 249–259. New York: Alan R. Liss.

Smith, Douglas W. and Gary Ferguson. 2005. *Decade of the Wolf: Returning the Wild to Yellowstone*. Guildford, CT: Lyons Press.

Smuts, Barbara. 1985. *Sex and Friendship in Baboons*. New York: Aldine Publishing.

———. 1992. Male aggression against women: An evolutionary perspective. *Human Nature* 3: 1–44.

———. 2000. Battle of the sexes. In Marc Bekoff, ed., *The Smile of a Dolphin*, 92–95. New York: Discovery Books.

Smuts, Barbara and John M. Watanabe. 1990. Social relationships and ritualized greetings in adult male baboons (*Papio cynocephalus anubis*). *International Journal of Primatology* 11, 2: 147–172.

Sorin, Anna Bess. 2004. Paternity assignment for white-tailed deer (*Odocoileus virginianus*): Mating across age classes and multiple paternity. *Journal of Mammalogy* 85,2: 356–362.

Spinage, C.A. 1982. *A Territorial Antelope: The Uganda Waterbuck*. London: Academic Press.

Steinhart, Peter. 1995. *In the Company of Wolves*. New York: Knopf.

Strum, Shirley C. 1987. *Almost Human: A Journey into the World of Baboons*. New York: Random House.

Struhsaker, Thomas T. 1975. *The Red Colobus Monkey*. Chicago: University of Chicago Press.

———. 1977. Infanticide and social organization in the redtail monkey (*Cercopithecus ascanius schmidti*) in the Kibale Forest, Uganda. *Zeitschrift für Tierpsychologie* 45: 75–84.

Thomas, Elizabeth Marshall. 1993. *The Hidden Life of Dogs*. Boston: Houghton Mifflin.

———. 1994. *The Tribe of Tiger*. New York: Simon and Schuster.

Thompson, Ernest Seton. 1942. *Wild Animals I Have Known*. New York: Scribners. (Orig. pub. 1898, repr. 1926.)

Thornton, Alex and Katherine McAuliffe. 2006. Teaching in wild meerkats. *Science* 313,5784: 227–229.

Van Noordwijk, Maria A. and Carel van Schaik. 2001. Career moves: Transfer and rank challenge decisions by male long-tailed macaques. *Behaviour* 138,3: 359–395.

Veenema, Hans C., Berry M. Spruijt, Willem Hendrik Gispen, and Jan A.R.A.M. van Hooff. 1997. Aging, dominance history, and social behavior in Java-monkeys (*Macaca fascicularis*). *Neurobiology of Aging* 18,5: 509–515.

Veenema, Hans C., Jan A.R.A.M. van Hooff, Willem Hendrik Gispen, and Berry M. Spruijt. 2001. Increased rigidity with age in social behavior of Java-monkeys (*Macaca fascicularis*). *Neurobiology of Aging* 22: 273–281.

Venne, Sharon Helen. 1998. *Our Elders Understand Our Rights*. Penticton, BC: Theytus Books.

de Waal, Frans. 1982. *Chimpanzee Politics*. London: Jonathan Cape.

———. 1989. *Peacemaking among Primates*. Cambridge, MA: Harvard University Press.

———. 1996. *Good Natured: The Origins of Right and Wrong in Humans and Other Animals*. Cambridge, MA: Harvard University Press.

———. 1999. Cultural primatology comes of age. *Nature* 399: 635–636.

———. 2005. *Our Inner Ape: A Leading Primatologist Explains Why We Are Who We Are*. New York: Riverhead Books.

Walters, J.R. 1990. Red-cockaded woodpeckers: A "primitive" cooperative breeder. In Peter B. Stacey and Walter D. Koenig, eds., *Cooperative Breeding in Birds*, 67–101. Cambridge: Cambridge University Press.

Waterhouse, Mary L. 1983. A life-stage analysis of Taiwanese women: Social and health-seeking behaviors. In Samuel K. Wasser, ed., *Social Behavior of Female Vertebrates*, 215–232. New York: Academic Press.

Watson, Lyall. 2002. *Elephantoms: Tracking the Elephant*. New York: Norton.

Watson, Paul. 2006. Priorities in the plight of ocean creatures. *Vancouver Sun*, December 8, A9.

Webb, Betsy. 2007. *The Emotional Lives of Animals*. Novato, CA: New World Library.

Weber, Bill and Amy Vedder. 2001. *In the Kingdom of Gorillas*. New York: Simon and Schuster.

Weiner, Jonathan. 1994. *The Beak of the Finch*. New York: Vintage Books.

Weladji, Robert B., Atle Mysterud, Oystein Holand, and Dag Lenvik. 2002. Age-related reproductive effort in reindeer (*Rangifer tarandus*): Evidence of senescence. *Oecologia* 131,1: 79–82.

Wells, Randall S. 2003. Dolphin social complexity: Lessons from long-term study and life history. In Frans B.M. de Waal and Peter L. Tyack, eds., *Animal Social Complexity: Intelligence, Culture, and Individualized Societies,* 32–56. Cambridge, MA: Harvard University Press.

Werner, Tracey K. and Thomas W. Sherry. 1987. Behavioral feeding specialization in *Pinaroloxias inornata,* the "Darwin's Finch" of Cocos Island, Costa Rica. *Proceedings of the National Academy of Science* 84,15: 5506–5510.

Whitehead, Hal. 2003. *Sperm Whales: Social Evolution in the Ocean.* Chicago: University of Chicago Press.

Williams, Elma M. 1963. *Valley of Animals.* London: Hodder and Stoughton.

Winkler, David W. 2004. Nests, eggs and young: Breeding biology of birds. In Sandy Podulka, Ronald W. Rohrbaugh, and Rick Bonney, eds., *Handbook of Bird Biology,* 2nd ed., pt. 2, chapt. 8: 1–152. Ithaca, NY: Cornell University Press.

Wolfe, Linda D. and M.J. Sabra Noyes. 1981. Reproductive senescence among female Japanese macaques (*Macaca fuscata fuscata*). *Journal of Mammalogy* 62,4: 698–705.

Woodhouse, Barbara. 1954. *Talking to Animals 'The Woodhouse Way'.* Harmondsworth, UK: Penguin Books.

Woolfenden, G.E. and J. W. Fitzpatrick. 1990. Florida scrub jays: A synopsis after 18 years of study. In Peter B. Stacey and Walter D. Koenig, eds., *Cooperative Breeding in Birds,* 239–266. Cambridge: Cambridge University Press.

Yamagiwa, Juichi. 1987. Male life history and the social structure of wild mountain gorillas (*Gorilla gorilla beringei*). In Shoichi Kawano, Joseph H. Connell, and Toshitaka Hidaka, eds., *Evolution and Coadaptation in Biotic Communities,* 31–51. Tokyo: University of Tokyo Press.

Zahavi, A. 1990. Arabian babblers: The quest for social status in a cooperative breeder. In Peter B. Stacey and Walter D. Koenig, eds., *Cooperative Breeding in Birds,* 103–130. Cambridge: Cambridge University Press.